THE MODEL THEORY OF GROUPS

Notre Dame Mathematical Lectures
Number 11

The Model Theory of Groups

Ali Nesin and Anand Pillay, editors

University of Notre Dame Press
Notre Dame, Indiana

Library of Congress Cataloging-in-Publication Data

The Model theory of groups.

(Notre Dame mathematical lectures; 11)
1. Group theory. I. Nesin, Ali. II. Pillay, Anand. III. Series:
Notre Dame mathematical lectures; no. 11.
QA1.N87 no. 11a [QA171]   510 s   88-40322
ISBN 0-268-01371-3   [512'.22]

Manufactured in the United States of America

CONTENTS

Preface

This volume contains contributions by participants in the Stable Group seminar at Notre Dame. This seminar began during the 1985-86 Notre Dame Logic Year and continued up to December 1987.

Except for Simon Thomas' paper on expansions of projective spaces, all the papers here are connected in one way or another with stable group theory. This volume contains introductory, expository and research papers in the area and we hope will be of interest to beginners as well as "experts". We cover much of the material in Poizat's book Groupes Stables and go beyond it in some respects (for example Bouscaren's paper on Hrushovski's important group configuration theorem in which the presence of a definable group is recognized from a certain forking configuration). So we hope our volume will serve as a useful complement to Poizat's book.

We would like to thank the University of Notre Dame (Mathematics Department and College of Science), the National Science Foundation (Grant DMS 85 09920) and the Argonne Universities Association Trust Fund, for their support of the 1985-86 Logic Year at Notre Dame, which in addition to funding a conference in April 1986, allowed us to invite many of the contributors to this volume.

We would also like to thank Joan Hoerstman, Karen Jacobs and Leona Strickland for their excellent and patient work in preparing the manuscript.

Ali Nesin
Anand Pillay

Notre Dame, Indiana
October 1988

MODEL THEORY, STABILITY THEORY & STABLE GROUPS

Anand Pillay*

The aim of this chapter is to introduce the reader to the theory of stable groups not to give a rigorous exposition of the general theory. Thus we tend to proceed from the concrete to the abstract, with several examples and analyses of special cases along the way. On the other hand, getting to grips with stable groups presupposes some understanding of the point of view of model theory in general and stability theory in particular, and the first few sections are devoted to the latter.

1. MODEL THEORY

By a <u>relational structure</u> $\underline{M}$ we understand a set M (called the universe or underlying set of $\underline{M}$) equipped with relations R_i of arity $n_i < \omega$ say, for $i \in I$. Namely, for $i \in I$, R_i is a subset of the Cartesian product M^{n_i}. Here I and $<n_i : i \in I>$ depend on $\underline{M}$ and are called the <u>signature</u> of M. We also insist that I always contains a distinguished element $i_=$ such that $R_{i=}$ is the diagonal $\{(a,a) : a \in M\} \subseteq M^2$. Often the distinction between M and $\underline{M}$ is blurred. The model theorist is interested in certain subsets of M and of M^n (the <u>definable</u> sets) which are obtained in a simple fashion from the R_i. So $\mathcal{D}(M)$ is a collection of subsets of M^n, $n < \omega$, which can be characterized as follows:

(i) Every $R_i \in \mathcal{D}(M)$.

(ii) If $n < \omega$, $X \in \mathcal{D}(M)$ is a subset of M^n and π is a permutation of $\{1,...,n\}$ then $\pi(X) = \{(a_{\pi(1)},...,a_{\pi(n)}) : (a_1,...,a_n) \in X\} \in \mathcal{D}(M)$.

(iii) $\mathcal{D}(M)$ is closed under Boolean combinations, i.e. if $n < \omega$ and

* Partially supported by NSF Grant DMS 8601289

$X, Y \in \mathcal{D}(M)$ are subsets of M^n then $X \cup Y, X \cap Y, M^n - X$ are all in $\mathcal{D}(M)$.

(iv) If $X \in \mathcal{D}(M)$ and $Y \in \mathcal{D}(M)$ then $X \times Y \in \mathcal{D}(M)$.

(v) If $X \in \mathcal{D}(M)$ is a subset of M^{n+m}, then the projection of X on M^n is in $\mathcal{D}(M)$.

(vi) If $X \in \mathcal{D}(M)$ is a subset of M^{n+m} and $\bar{a} \in M^n$ then $X_{\bar{a}} = \{\bar{b} \in M^m: (\bar{a},\bar{b}) \in X\}$ is in $\mathcal{D}(M)$.

(vii) Nothing else is in $\mathcal{D}(M)$.

We call $\mathcal{D}(M)$ the class of definable sets of M.

These definable sets can be defined (and usually are) syntactically. Associated to the relational structure M (in fact to its signature) is a language L(M) consisting of symbols: P_i for each $i \in I$, "variables" x_j for each $j < \omega$, and logical symbols $\wedge$ (and), $\vee$ (or), $\neg$ (not), $\forall$ (for all) and $\exists$ (there exists). L(M) -formulas are constructed from these symbols as follows: if x_j are variables then $P_i x_1 \ldots x_{n_i}$ is an (atomic) formula. If φ, ψ are formulas and x is a variable then $\varphi \wedge \psi$, $\varphi \vee \psi$, $\neg \varphi$, $(\exists x)\varphi$, $(\forall x)\varphi$ are all formulas. A variable x is said to be free in the formula φ if some occurrence of x in φ is not in the scope of any quantifier. We write $\varphi(x_1,\ldots,x_n)$ to mean that $x_1,\ldots,x_n$ are the free variables in the formula φ. We then define "$\varphi(x_1,\ldots,x_n)$ is true of $(a_1,\ldots,a_n)$ in M" (where $a_1,\ldots,a_n \in M$) as follows:

If φ is atomic, say $Py_1 \ldots y_n$, and for some permutation π of $\{1,\ldots,n\}$, $x_i = y_{\pi(i)}$ then $\varphi(x_1,\ldots,x_n)$ is true of $(a_1,\ldots,a_n)$ in M if $(a_{\sigma(1)},\ldots,a_{\sigma(n)}) \in P$ where $\sigma = \pi^{-1}$.

If ψ is $(\exists x_{n+1})\,\varphi$ and x_{n+1} is a free variable of φ, then $\psi(x_1,\ldots,x_n)$ is true of $(a_1,\ldots,a_n)$ in M if there is $a_{n+1} \in M$ such that $\varphi(x_1,\ldots,x_n,x_{n+1})$ is true of $(a_1,\ldots,a_{n+1})$ in M. Similarly for $\psi = (\forall x_{n+1})\,\varphi$.

The clauses for $\wedge, \vee, \neg$ are obvious. We abbreviate "$\varphi(x_1,\ldots,x_n)$ is true of $(a_1,\ldots,a_n)$ in M" by the notation $M \models \varphi(a_1,\ldots,a_n)$. (Note this notation depends on our having listed the free variables in φ in a certain order).

By abuse of everything, we can and will think of $M \models \varphi(a_1,\ldots,a_n)$ as saying that φ is true when we substitute a_i for x_i.

It is now routine to check that

Fact 1.1. If $X \subset M^n$, then $X \in \mathcal{D}(M)$ if and only if there are an L(M) formula $\varphi(x_1,...,x_n,y_1,...,y_m)$ and $b_1,...,b_m \in M$ such that

$$X = \{\bar{a} \in M^n\colon M \models \varphi(\bar{a}, \bar{b})\}.$$

The syntactic approach to defining definable sets appears at first to be preferable as one can make the following definition.

Definition 1.2. $X \in \mathcal{D}(M)$, a subset of M^n, is said to be A-definable or defined over A (for $A \subset M$) if in Fact 1.1 we can choose φ with $\bar{b} \subset A$.

Example 1.3. Let K be an algebraically closed field. We can consider K as a relational structure in the above sense by choosing {0}, {1}, and the graphs of addition and multiplication as the "distinguished" relations. Note that if $P_1,...,P_r$ are polynomials in n-variables over K, then the subset V of K^n consisting of the simultaneous zero set of these polynomials is a definable set. These are called the affine algebraic sets. Finite Boolean combinations of such sets are called constructible sets of K, and either Tarski's "quantifier elimination theorem" (quantifier elimination in a language with function symbols for addition and multiplication) or Chevalley's theorem states

Fact: The constructible sets of K are precisely the definable sets of K.

For an affine algebraic set $X \subseteq K^n$ there is an algebraic-geometrical notion of X being defined over k (k a subfield of K) which may have some discrepancy with the model theoretic notion (Definition 2.2). Namely: let I(X) $\subseteq K[x_1,...,x_n]$ be the ideal of polynomials which vanish on X. According to the algebraic geometer X is defined over $k \subset K$ if I(X) can be generated as an ideal by polynomials in $k[x_1,...,x_n]$.

We do have (for $X \subseteq K^n$ affine algebraic and k subfield of K)

Fact: X is defined over k in the model theoretic sense iff X is defined over $k^{p^{-\infty}}$ in the sense of algebraic geometry (where p = char K).

So if k is perfect, or char k = 0, the notions agree.

1.4. The usual procedure in model theory is to start with a language L and to consider various subclasses of L-structures. So L will essentially be a signature as above, i.e. will consist of a set of relation symbols of specified arity and an L-structure will be a relational structure equipped with corresponding relations of the right arity. This enables us to compare L-structures in various respects. For instance, by an L-sentence we mean an L- formula which has no free variables. An L-structure M is said to be a model of a set Γ of L-sentences if for every $\sigma \in \Gamma$, $M \models \sigma$, i.e. every $\sigma \in \Gamma$ is true in M. A set of L-sentences Γ is said to be consistent if it has a model. A consistent set of sentences Γ is said to be a complete theory if for every L-sentence σ either $\sigma \in \Gamma$ or $\neg\sigma \in \Gamma$, equivalently for some M, $\Gamma = \{\sigma: M \models \sigma\}$; in the latter case Γ being called the theory of M. Two L-structures M and N are called elementarily equivalent if they have the same theory, equivalently they satisfy the same L-sentences. As an example, any two algebraically closed fields of the same characteristic, say p, are elementarily equivalent; in other words the set T_{ACF_p} of sentences (in the language in Example 1.3 for example) true in all algebraically closed fields of characteristic p is a complete theory.

A crucial tool in model theory is the compactness theorem: a set of sentences Γ is consistent iff every finite subset of Γ is consistent. This gives substance to the following important notion: Let M, N be L-structures with M a substructure of N ($M \subseteq N$, with the obvious meaning). M is said to be an elementary substructure of N, $M \prec N$, if for every formula $\varphi(\bar{x})$ of L and $\bar{a} \in M$, we have $M \models \varphi(\bar{a})$ iff $N \models \varphi(\bar{a})$.

Let us remark that if $M \prec N$ then any definable set $X \subseteq M^n$ in M has a canonical extension to a definable set $X' \subseteq N^n$ in N. Namely, let $\varphi(\bar{x},\bar{y})$, $\bar{a} \subseteq M$ be such that $\varphi(\bar{x},\bar{a})$ defines X in M. Then let $X' = \{\bar{x} \in N^n : N \models \varphi(\bar{x},\bar{a})\}$. Note that $X \subseteq X'$ and X' does not depend on the particular choice of φ and $\bar{a}$.

The compactness theorem yields for any infinite M, elementary extensions N of M of arbitrarily large cardinality. Another consequence of

Tarski's quantifier elimination is that if $K_1 \subseteq K_2$ are algebraically closed fields then $K_1 \prec K_2$, noting the following characterisation : let $M_1 \subseteq M_2$, then $M_1 \prec M_2$ iff for any non-empty M_1-definable subset X of M_2, $X \cap M_1 \neq \emptyset$.

1.5. Saturated models.

Let κ be an infinite cardinal. The structure N is said to be <u>κ-saturated</u> if for any $A \subset N$ with $|A| < \kappa$ and any collection $X_i, i \in I$ of A-definable subsets of N with the finite intersection property ($\bigcap_{i \in J} X_i \neq \emptyset$ for all finite $J \subseteq I$), we have $\bigcap_{i \in I} X_i \neq \emptyset$. Again the compactness theorem gives for any M and κ some κ-saturated $N \succ M$.

It is worth noting that the definition above of κ-saturation would be equivalent if we allowed the X_i to be A-definable subsets of N^n for any $n \geq 1$ This apparently stronger fact follows by use of the <u>existential quantifier</u>.

One can think of the property of κ-saturation of N as meaning that for any $M \prec N$ with $|M| < \kappa$, any situation that can happen in some elementary extension of M already happens in N. (In this sense N is like a universal domain. In fact, what Weil calls a universal domain - an algebraically closed field of infinite transendence degree κ over the prime field- is κ-saturated). Moreover if $M \equiv N$ and $|M| < \kappa$ then there is an elementary embedding (obvious meaning) of M into N. <u>It will be convenient to assume that any complete theory has models which are κ-saturated and of cardinality κ, for arbitrarily large κ</u>. Such a model, N say, will have homogeneity properties in addition to saturation properties, which are pointed out subsequently. (For stable theories the existence of such models is guaranteed. Otherwise, it depends on set theory).

Let us now fix such a model N (κ-saturated of cardinality κ for some large κ). A, A′, B etc. will denote subsets of N of cardinality $<\kappa$, and M, M′, M_1,... elementary substructures of N of cardinality $< \kappa$ (often called <u>models</u>). We now introduce the important notion of a <u>type</u>.

Let $A \subset N$. By a complete n-type over A we mean a maximal consistent collection of A-definable subsets of N^n (where consistent means having the finite intersection property). Alternatively, with some abuse of earlier notation, a complete n-type over A is a maximal set Γ of formulas of the form $\varphi(x_1,...,x_n,\bar{a})$ where $\bar{a} \subset A$ and for $\varphi_1,...,\varphi_m \in \Gamma$, $N \models \exists \bar{x} (\bigwedge_{i=1}^{m} \varphi_i(\bar{x}))$.

Let $b_1,...,b_n \in N$. By the type of $\bar{b}$ over A (in N if you wish), $tp(\bar{b}/A)$ is meant the collection of A-definable subsets of N^n containing $\bar{b}$. $tp(\bar{b}/A)$ is clearly a complete n-type over A. Conversely, saturation of N implies that every complete n-type Γ over A is the form $tp(\bar{b}/A)$, for some $\bar{b} \in N^n$. $\bar{b}$ is said to realize Γ. The set of complete n-types over A is denoted $S_n(A)$, and types themselves are usually denoted by p, q etc.

The fact that N is saturated in its own cardinality gives us a nice characterization: if $\bar{b}_1 \in N^n$, $\bar{b}_2 \in N^n$ then $tp(\bar{b}_1/A) = tp(\bar{b}_2/A)$ iff there is an automorphism f of N such that $f(\bar{b}_1) = \bar{b}_2$ and f fixes A pointwise. (Similarly for types of infinite tuples of cardinality $< \kappa$).

Saturation of N also enables us to give the notion "definable over N" a "Galois theoretic" interpretation. Firstly, the compactness theorem yields: Let $X \subseteq N^n$ be definable, let $A \subset N$ and suppose that whether or not some $\bar{b} \in N^n$ is in X depends only on $tp(\bar{b}/A)$. Then X is A-definable. In conjunction with the previous observation this shows that for definable $X \subseteq N^n$, X is A-definable iff for every automorphism f of N which fixes A pointwise, $f(X) = X$.

1.6. N^{eq}

It will be sometimes convenient (especially when dealing with groups) to work in a structure which is "closed under definable quotients". We can construct from N such a universe, N^{eq}, which is "essentially" the same as N. Informally, N^{eq} is the disjoint union of a collection of universes, one of which is N, and each being picked out by a new predicate. Each new universe is identified, by means of a new function symbol, with the set of classes of a

$\emptyset$–definable equivalence relation on N^m for some $m < \omega$. Moreover, all $\emptyset$-definable equivalence relations on N^m, $m < \omega$ are accounted for by these new universes. More formally, if L is the original language then L_{eq} will be L augmented by a unary predicate symbol P_E and a function symbol f_E for every $\emptyset$-definable equivalence relation E. N^{eq} will be the L_{eq} structure whose underlying set is the disjoint union of the interpretations of the various P_E's (which we can call N_E). $N_=$ is precisely N with its original L-structure. If E is an equivalence relation on N^m, then the interpretation of f_E is a surjective map $N^m \to N_E$ whose fibres are the E-classes. This whole construction is of course a function of Th(N), so for any $M \equiv N$, we can obtain in the same way M^{eq}. We however make the additional stipulation that in the models of T^{eq} (= Th(N^{eq})), every element should satisfy one of the predicates P_E (So T^{eq} is a "many sorted" theory). This can be formally accomplished by requiring that every L_{eq} formula we consider must state for each of its variables the predicate P_E in which the variable lies. We think of the P_E's as picking out certain sorts.

N^{eq} has the following properties:

1.7. (i) A subset X of N^m definable in N^{eq} is definable in N.

(ii) Any automorphism of N has a unique extension to an automorphism of N^{eq}.

(iii) For any definable subset X of $(N^{eq})^n$ there is an element $a_X \in N^{eq}$ such that an automorphism f of N^{eq} fixes X setwise iff it fixes a_X.

An element b is said to be <u>algebraic</u> over A ($A, b \subset N$ or even N^{eq}) if b lies in some finite A-definable set. b is <u>definable</u> over A if $\{b\}$ is A-definable.

acl (A) = {b: b is algebraic over A},

dcl (A) = {b: b definable over A}.

A definable set X is said to be <u>almost over</u> A (X, A in N or N^{eq}) if X has finitely many images under A-automorphisms of N.

Fact 1.8. X is almost over A iff X is acl(A)-definable in N^{eq}.

2. ω-STABILITY

Stability is an hypothesis on the "complexity" of the family of definable sets in a model. Generally we talk of stability, superstability, ω-stability (or total transcendence) of a <u>complete theory T</u>, which translates into certain rank or dimension functions on the definable sets of a saturated model N of T being everywhere defined. Probably the easiest such rank to define and understand is Morley rank, RM.

Definition 2.1. Let $n < \omega$, $X \subseteq N^n$ a definable set. $RM_n(X)$ is defined as follows:

(i) $RM_n(X) \geq 0$ if $X \neq \varnothing$.
$RM_n(X) \geq \delta$ if $RM_n(X) \geq \alpha$ for all $\alpha < \delta$ (δ limit).
$RM_n(X) \geq \alpha + 1$ if there are pairwise disjoint definable subsets $X_i \subset N^n$ for $i < \omega$ such that $RM_n(X \cap X_i) \geq \alpha$ for all $i < \omega$. If $RM_n(X) = \alpha$ some α, we say $RM_n(X)$ is defined. Otherwise (i.e. if $RM_n(X) \geq \alpha$, for all α) we put $RM_n(X) = \infty$.

(ii) If $p \in S_n(A)$, $A \subset N$, we put $RM_n(p) = \min \{RM_n(X): X \in p\}$.

Remarks and Definitions 2.2.

(i) Let M be an arbitrary structure. Let N be a κ-saturated elementary extension of M of cardinality κ, where $\kappa > |M| +$ cardinality of $L(M)$. Let $X \subseteq M^n$ be definable in M. Let $X' \subseteq N^n$ be the canonical extension of X to a set definable in N (as in 1.4). Then we define $RM_n(X) = RM_n(X')$.

(ii) Let T be a complete theory and let N be a model of T (N saturated in its own cardinality κ, for large κ). We say T is <u>totally transcendental</u> if for every definable $X \subseteq N^n$, $RM_n(X)$ is defined (actually it is enough to demand this for $X \subseteq N$).

(iii) If the language of T is countable and T is totally transcendental, we say that T is ω-stable. (This notation will be explained later: basically here it means that for countable models M of T, $S_1(M)$ is countable). We may, by abuse of language, still use the expression ω-stable to mean totally transcendental, even if T is not countable.

(iv) We usually drop the subscript n from RM_n when the arity of X is either clear from the context or unimportant.

Fact 2.3.

(i) If $RM(X) = \alpha$ and X is A-definable then there is a complete type $p \in S(A)$, with $RM(p) = \alpha$ and $X \in p$.

(ii) If $RM(X) = \alpha$ then there is a greatest $k < \omega$ for which there are pairwise disjoint X_i for $i = 0,1,\ldots,k-1$ such that $RM(X \wedge X_i) = \alpha$ $\forall i < k$. k is called the Morley degree of X. Similarly one can define the Morley degree of a complete type $p \in S(A)$.

Let us remark that if we work in N^{eq} then for any sort S (S is one of the P_E), we can define $RM_S(X)$ for X a definable set of elements of sort S. It will then be the case that if RM(X) is defined for all $X \subseteq N$ then also $RM_S(X)$ is defined for every sort S in N^{eq} (i.e. T totally transcendental $\Rightarrow$ T^{eq} is totally transcendental).

Example 2.4. Strongly minimal sets

Let $X \subseteq N^n$ be a definable set in (saturated) N. X is said to be strongly minimal if for every definable $Y \subseteq N^n$ either $X \cap Y$ or $X - Y$ is finite.

Similarly, working in N^{eq} we can speak of a definable set X in sort S being strongly minimal. The complete theory T is said to be strongly minimal if the universe of a saturated model N of T is strongly minimal (Note: it does not make sense to speak of T^{eq} being strongly minimal). As a case study we will show what Morley rank means in the case of strongly minimal theories (and also strongly minimal sets).

We first make a trivial remark:

Remark: Let $b, a_1,\dots,a_k \in N$, $A \subset N$ with $b \in \mathrm{acl}(a_1,\dots,a_k,A)$. Then there is an A-definable set $X \subseteq N^{k+1}$ such that $(a_1,\dots,a_k, b) \in X$ and such that whenever $(a_1',\dots,a_k',b') \in X$ then $b' \in \mathrm{acl}\,(a_1',\dots,a_k',A)$.
We will say that the set $\{a_1,\dots,a_k\}$ is <u>algebraically independent</u> over A if for all i, $a_i \notin \mathrm{acl}(a_1,\dots,a_{i-1},a_{i+1},\dots,a_k,A)$.

(i) of the next fact is easy, but (ii) is less so.

Fact 2.5. Let N be strongly minimal, $A \subset N$.

(i) Let $a_1,a_2 \in N$, $a_1,a_2 \notin \mathrm{acl}\,(A)$. Then $\mathrm{tp}(a_1/A) = \mathrm{tp}(a_2/A)$.

(ii) Let $a_1,\dots,a_k \in N$. Then all maximal algebraically independent over A subsets of $\{a_1,\dots,a_k\}$ have the same size m. We call this number $\dim(a_1,\dots,a_k / A)$. This being clearly a function of $\mathrm{tp}(\bar{a}/A)$ (where $\bar{a} = (a_1,\dots,a_k)$) we write also $m = \dim\,(\mathrm{tp}(\bar{a}/A))$.

On the other hand, let $X \subseteq N^k$ be defined over A. We define $\dim X$ to be $\max\,\{\dim\,(\bar{a}/A) : \bar{a} \in X\}$. This does not depend on A. Namely suppose $\dim X$ as defined above is equal to r. Let $B \supset A$. Let $\dim\,(\bar{a}/A) = r$ where $\bar{a} = (a_1,\dots,a_r,\dots,a_k) \in X$ and without loss of generality $\{a_1,\dots,a_r\}$ is algebraically independent over A. Let $\{b_1,\dots,b_r\}$ be algebraically independent over B. By Fact 2.5 (i) applied repeatedly, $\mathrm{tp}(a_1,\dots,a_r/A) = \mathrm{tp}(b_1,\dots,b_r/A)$. Thus by saturation of N we can extend $(b_1,\dots,b_r)$ to a sequence $\bar{b} \in N^k$ such that $\mathrm{tp}(\bar{b}/A) = \mathrm{tp}(\bar{a}/A)$. In particular $\bar{b} \in X$ and clearly $\dim(\bar{b}/B) = r = \max\,\{\dim\,(\bar{b}' / B) : \bar{b}' \in X\}$

We aim to show that for strongly minimal N, Morley rank equals dimension for types or definable sets of k-tuples, $k < \omega$.

Lemma 2.6. (N strongly minimal). Let $X \subset N^k$ be definable. Then $\dim X \geq r + 1$ iff there are pairwise disjoint definable $X_i \subseteq X$ for $i \in \omega$ such that $\dim X_i \geq r$ for all r.

Proof: Note first that every nonempty definable set has dimension ≥ 0. Suppose now $\dim X \geq r+1$. Suppose X to be A-definable and let $(a_1,...,a_{r+1},...,a_k) = \bar{a} \in X$ with $\dim(\bar{a}/A) \geq r+1$. Without loss of generality $\{a_1,...,a_{r+1}\}$ is algebraically independent over A. Let $\{b_1^i : i < \omega\}$ be algebraically independent over A (by saturation of N). By Fact 2.5. $tp(b_1^i/A) = tp(a_1/A)\ \forall\, i < \omega$, thus by "homogenity" of N there are $\bar{b}^i = (b_1^i,...,b_k^i)$ for $i < \omega$ such that $tp(\bar{b}^i/A) = tp(\bar{a}/A)$. For $i < \omega$ let $X_i = \{\bar{x} : \bar{x} \in X \text{ and } x_1 = b_1^i\}$. Then $X_i \subset X$, the sets X_i are pairwise disjoint and $\dim X_i \geq r$ (for the latter, note X_i is $A \cup \{b_1^i\}$ -definable, $\bar{b}^i \in X_i$ and $\{b_2^i,...,b_{r+1}^i\}$ is algebraically independent over $A \cup \{b_1^i\}$). So this shows left to right.

Conversely, suppose that $X_i \subseteq X$ for $i \in \omega$, with the X_i pairwise disjoint and $\dim X_i \geq r$. Suppose X and all the X_i to be B-definable. For each i let $\bar{b}^i = (b_1^i,...,b_k^i)$ be in X_i with $\dim(\bar{b}^i/B) \geq r$. As there are infinitely many i, we can assume that for each i, $\{b_1^i,...,b_r^i\}$ is algebraically independent over B. By repeated application of Fact 25 (i), $tp(b_1^i,...,b_r^i/B) = tp(b_1^j,...,b_r^j/B)$ for $i, j < \omega$. So by section 1.4 for each $i, j < \omega$ there is a B-automorphism taking $(b_1^i,...,b_r^i)$ to $(b_1^j,...,b_r^j)$. As every B-automorphism leaves each X_i as well as X setwise invariant, we can assume that there are $b_1,...,b_r$ such that for all $i < \omega$ and $j \leq r$, $b_j^i = b_j$ (namely the $\bar{b}^i$ have same first r coordinates).

Claim: $Y = \{\bar{c} \in N^{k-r} : (b_1,...,b_r, \bar{c}) \in X\}$ is infinite. For otherwise, as $X_i \subseteq X$ for $i < \omega$, there would be $\bar{c}$ with $(b_1,...,b_r, \bar{c}) \in X_i \cap X_j$ for some $i \neq j$, contradicting pairwise disjointness.

By the claim and saturation of N we can find $\bar{c} \in Y$ and some coordinate of $\bar{c}$, say c_1 such that $c_1 \notin acl(b_1,...,b_r,B)$. But then

$\dim X \geq r+1$. This completes the proof of Lemma 2.6. □

Corollary 2.7. (N strongly minimal).

(i) Let $X \subseteq N^k$ be definable, then $\dim X = RM(X)$.

(ii) Let $a_1,...,a_k \in N$, $A \subset N$, Then $\dim(a_1,...,a_k/A) = RM(tp(a_1,...,a_k/A))$.

Proof: (i) follows from Lemma 2.6 and the definition of Morley rank. (ii) follows by noting that if $p = tp(a_1,...,a_k/A)$ then $\dim p = \min\{\dim X : X \in p\} = RM(p)$. □

Example 2.8. Let K be a saturated algebraically closed field (with no additional structure beyond the field structure).

Lemma 2.9. (i) K is strongly minimal.

(ii) If $k \subset K$ and $a_1,...,a_n$ are in K, then $a_1,...,a_n$ are algebraically independent over k in the sense of model theory iff they are algebraically independent over k in the sense of field theory.

Proof: (i) We already remarked in Example 1.3 that any definable set $X \subset K$ is constructible, i.e. a finite Boolean combination of algebraic subsets of K. Noting that an algebraic subset of K must be finite (or all of K), we see that X is finite or cofinite, whereby K is strongly minimal.

(ii) We use again the fact (quantifier elimination) that every formula $\varphi(\overline{x})$ is equivalent in K to a quantifier free formula $\psi(\overline{x})$ in a language with function symbols for $+, \cdot$ and constant symbols for 0, 1. It easily follows that if $a_1,...,a_n$ are algebraically dependent over k in the model-theoretic sense then $a_1,...,a_n$ satisfy a nontrival polynomial relation over k. □

Let now $V = V(\overline{a}/k) = \{\overline{b} \in K^n : f(\overline{b}) = 0$ whenever $f(\overline{a}) = 0$ for $f \in k[\overline{x}]\}$. By definition, the <u>algebraic-geometrical dimension</u> of this affine algebraic set V is the transcendence degree of $k(\overline{a})$ over k.

Proposition 2.10. (With the above notation)

RM(tp($\bar{a}$/k)) = RM(V) = algebraic geometrical dimension of V.

Proof: By part (ii) of the Lemma

(a) transcendence degree of k($\bar{a}$) over k = dim ($\bar{a}$/k).

By Corollary 2.7 (ii)

(b) RMtp($\bar{a}$/k) = dim ($\bar{a}$/k).

By part (ii) of the Lemma again, dim ($\bar{b}$/K) $\leq$ dim ($\bar{a}$/K) for all $\bar{b} \in V$ and thus

(c) dim(V) = dim ($\bar{a}$/k).

The Proposition now follows from (a), (b) and (c). □

3. STABILITY

Although much of this volume will concentrate on ω-stable groups, and even ω-stable groups of finite Morley rank, it is worth saying something about stable theories in general. Stability is a property of certain theories (the stable ones) which is considerably weaker than ω-stability. It might be considered as "local ω-stability", and we subsequently introduce it in this way. For now, we can take a definition of stability as: T is stable if there is no formula $\varphi(\bar{x}, \bar{y})$ and tuples $\bar{a}_i, \bar{b}_j$ $(i,j < \omega)$ in a model M of T with $M \models \varphi(\bar{a}_i, \bar{b}_j)$ iff $i \leq j$; or equivalently there is no $\psi(\bar{x}_1, \bar{x}_2)$ and $\bar{a}_i$ $(i \leq \omega)$ with $M \models \psi(\bar{a}_i, \bar{a}_j)$ iff $i \leq j$. ("One cannot define an order").

Under the sole assumption of stability, a good notion of independence can be defined: for a model N of stable T, $\bar{a} \subset N$, $A \subset B \subset N$, we will make sense of " $\bar{a}$ is free from B over A", " $\bar{a}$ and B are independent over A". This will depend only on the formulas true of $\bar{a}$, A, B and so we will also say "tp($\bar{a}$/B) <u>does not fork over</u> A". For T ω-stable this will agree with "RM($\bar{a}$/B) = RM($\bar{a}$/A)". In general, the following will be true:

(i) $\bar{a}$ is free from B over A iff $\bar{a}$ is free from $B_o \cup A$ over A for every finite $B_o \subset B$.

(ii) For any $\bar{a}$, B there is $B_0 \subset B$ $|B_0| \leq |T|$, with $\bar{a}$ free from B over B_0.

(iii) If $A \subset B \subset C$, then $\bar{a}$ is free from C over A iff $\bar{a}$ is free from C over B and $\bar{a}$ is free from B over A.

(iv) $\bar{a}$ is free from B over A iff for all $\bar{b} \subset B$, $\bar{b}$ is free from $\bar{a} \cup A$ over A.

(v) For given $p \in S(A)$ and $B \supset A$ there are at most $2^{|T|}$ and at least one $q \in S(B)$ such that $q \supseteq p$ and q does not fork over A.

(vi) $\bar{a}$ is free from B over M (M a model $M \subset B$) iff $tp(\bar{a}/B)$ is <u>definable</u> over M i.e. for every $\psi(\bar{x}, \bar{y}) \in L$ there is $\delta(\bar{y})$ with parameters in M such that for all $\bar{b} \in B$, $N \models \delta(\bar{b})$ iff $N \models \varphi(\bar{a}, \bar{b})$.

We will also say $\bar{a}$ is free from A over A, whereby, by (vi) we see, every $p \in S(M)$ is definable.

There are a number of ways of introducing forking and proving its properties, the original being due to Shelah [S] and another influential treatment being due to Lascar and Poizat [L.P]. See also [P], [H.H], [R].

One of the more efficient procedures appears in the introduction of Hrushovski's thesis which is apparently the content of a course given at Berkeley by Harrington. For the interested reader we outline this approach giving selective proofs.

We work in a saturated model $\mathbb{C}$ of T. Let $\Delta(\bar{x})$ be a (usually finite) collection of L–formulas, say $\{\delta_i(\bar{x}, \bar{y}_i) : i \in I\}$ where $\bar{x} = (x_1,\ldots,x_n)$. By a Δ-type we mean a consistent (small) collection of formulas of the form $\delta(\bar{x}, \bar{a})$, $\neg\, \delta(\bar{x}, \bar{a})$, for $\delta(\bar{x}, \bar{y}) \in \Delta$ and $\bar{a} \subset \mathbb{C}$. A <u>complete</u> Δ-type over $A \subset \mathbb{C}$ is a Δ-type, all of whose formulas have parameters in A and which is maximal (consistent) such. Sometimes we identify a complete Δ-type over A with its closure under conjunctions and disjunctions. $S_\Delta(A)$ denotes the complete Δ-types over A.

Definition 3.1. A Δ-defining schema over A is a map that assigns to each $\delta(\bar{x}, \bar{y}) \in \Delta$ a formula $\psi_\delta(\bar{y})$ over A such that for any $B \supset A$ (or for some $B \supset A$ which is a model), the following set
$\{\delta(\bar{x},\bar{b}): \bar{b} \subset B, \delta \in \Delta, \models \psi_\delta(\bar{b})\} \cup \{\neg\delta(\bar{x}, \bar{b}) : \bar{b} \subset B, \delta \in \Delta, \models \neg \psi_\delta(\bar{b})\}$
is a complete Δ-type over B. If we let d denote the map $\delta(\bar{x}, \bar{y}) \rightarrow \psi_\delta(\bar{y})$ then we call the above complete Δ-type d(B). If $\Delta = L$, we just talk about a defining schema (over A).

We can observe immediately:

Lemma 3.2. (T stable). Let $p(\bar{x}) \in S(A)$, $q(\bar{y}) \in S(A)$ and $\varphi(\bar{x}, \bar{y}) \in L$. Let $\Delta_1(\bar{x})$ contain $\varphi(\bar{x}, \bar{y})$ and $\Delta_2(\bar{y})$ contains $\varphi(\bar{x}, \bar{y})$. Let d_1 be a Δ_1-defining schema over A such that for some (any) $M \supset A$, $p(\bar{x}) \cup d_1(M)$ is consistent. Let d_2 be a Δ_2-defining scheme over A such that for (any) $M \supset A$, $q(\bar{y}) \cup d_2(M)$ is consistent. Let $\psi_1(\bar{y})$ be $d_1(\varphi(\bar{x}, \bar{y}))$, and $\psi_2(\bar{x})$ be $d_2(\varphi(\bar{x},\bar{y}))$. Then $\psi_1(\bar{y}) \in q(\bar{y})$ if and only if $\psi_2(\bar{x}) \in p(\bar{x})$.

Proof: Suppose not and let $M \supset A$ be saturated. Without loss of generality, $\psi_2(\bar{x}) \in p(\bar{x})$ but $\neg \psi_1(\bar{y}) \in q(\bar{y})$. We define $\bar{a}_i$, $\bar{b}_i$ in M, inductively as follows: $\bar{a}_n$ realizes $p(\bar{x}) \cup d_1(A \cup \{\bar{a}_0, \bar{b}_0,\ldots,\bar{a}_{n-1}, \bar{b}_{n-1}\})$, and $\bar{b}_n$ realizes $q(\bar{y}) \cup d_2(A \cup \{\bar{a}_0,\bar{b}_0,\ldots,\bar{a}_{n-1},\bar{b}_{n-1},\bar{a}_n\})$. It is easy to check that $M \models \varphi(\bar{a}_i,\bar{b}_j)$ iff $i < j$, contradicting stability. □

The aim now is to show that for every $p(\bar{x}) \in S(A)$, where A is algebraically closed in $\mathbb{C}^{eq}$ and finite $\Delta(\bar{x})$, there is a Δ-defining schema d over A such that $p(\bar{x}) \cup d(M)$ is consistent, for all $M \supseteq A$. For this we introduce Δ-rank. So fix finite $\Delta(\bar{x})$. Let $\varphi(\bar{x})$ be a formula maybe with parameters.

Definition 3.3.

(i) $R_\Delta(\varphi(\bar{x})) \geq 0$ if $\models \exists \bar{x}\, \varphi(\bar{x})$.

(ii) $R_\Delta(\varphi(\overline{x})) \geq \alpha + 1$ if for every $m < \omega$ there are finite pairwise contradictory Δ-types $q_1,\ldots,q_m$ (i.e. for $i \neq j$ there is a formula in q_i whose negation is in q_j) such that $R_\Delta(\varphi(\overline{x}) \wedge \bigwedge q_i) \geq \alpha$, for $i = 1,\ldots,m$.

(iii) For limit δ, $R_\Delta(\varphi(\overline{x})) \geq \delta$ if $R_\Delta(\varphi(\overline{x})) \geq \alpha$ all $\alpha < \delta$.

(iv) If $\Phi(\overline{x})$ is a set of formulas then $R_\Delta(\Phi) = \min\{R_\Delta(\varphi(\overline{x})): \varphi$ is finite conjunction of members of $\Phi\}$.

(v) Suppose $R_\Delta(\Phi(\overline{x})) = \alpha$ and $m < \omega$ is greatest such that there are $q_1,\ldots,q_m$ as in (ii). We call m the Δ-multiplicity of $\varphi(\overline{x})$ and we denote it by $m_\Delta(\varphi(\overline{x}))$.

Fact 3.4. T is stable implies that for all $\varphi(\overline{x})$ and finite $\Delta(\overline{x})$, $R_\Delta(\varphi(\overline{x}))$ is defined (i.e. has an ordinal value).

Let us assume from now on T to be stable. We fix <u>finite</u> $\Delta(\overline{x})$, and we denote $(R_\Delta(\varphi(\overline{x}), m_\Delta(\varphi(\overline{x}))$ by $R\text{-}m_\Delta(\varphi(\overline{x}))$. We equip these pairs with the lexicographic ordering.

Fact 3.5.

(i) $R_\Delta(\varphi(\overline{x}) \vee \psi(\overline{x})) = \max\{R_\Delta(\varphi(\overline{x})), R_\Delta(\psi(\overline{x}))\}$.

(ii) Let $m > 1$, then $R\text{-}m_\Delta(\varphi(\overline{x})) \geq (\alpha,m)$ iff there are m_1,m_2 with $m_1 + m_2 = m$, $\delta(\overline{x},\overline{y}) \in \Delta$, $\overline{a} \in \mathbb{C}$ such that $R\text{-}m_\Delta(\varphi(\overline{x}) \wedge \delta(\overline{x},\overline{a})) \geq (\alpha,m_1)$ and $R\text{-}m_\Delta(\varphi(\overline{x}) \wedge \neg\, \delta(\overline{x},\overline{a})) \geq (\alpha,m_2)$.

(iii) Let $\Phi(\overline{x})$ be a consistent collection of formulas over A. Then there is a complete Δ-type $q(\overline{x})$ over A such that $R_\Delta(\Phi) = R_\Delta(\Phi \cup q)$.

Lemma 3.6. Let $R\text{-}m_\Delta(\varphi(\overline{x})) = (\alpha,m)$. Let $\delta(\overline{x},\overline{y}) \in \Delta$. Then there is a formula $\psi(\overline{y})$ (with parameters from among those in φ) such that for any $\overline{b}$

$$\mathbb{C} \models \varphi(\overline{b}) \text{ iff } R\text{-}m_\Delta(\varphi(\overline{x}) \wedge \delta(\overline{x},\overline{b})) = (\alpha,m).$$

Outline of proof: First, using Fact 3.5 (ii), the finiteness of Δ and induction one shows that for any formula $\varphi(\bar{x},\bar{y})$ and pair (α,m) there is a collection $\Gamma(\bar{y})$ of L-formulas such that $\mathbb{C} \models \Gamma(\bar{b})$ iff $\text{R-m}_\Delta(\varphi(\bar{x},\bar{b})) \geq (\alpha,m)$. Now, suppose $\text{R-m}_\Delta(\varphi(\bar{x})) = (\alpha,m)$. By Fact 3.5 (ii) "$\text{R-m}_\Delta(\varphi(\bar{x}) \wedge \delta(\bar{x},\bar{y})) \geq (\alpha,m)$" and "$\text{R-m}_\Delta(\varphi(\bar{x}) \wedge \neg\, \delta(\bar{x},\bar{y})) \geq (\alpha,1)$" are inconsistent. By compactness there is a formula $\psi(\bar{y})$ (with parameters among those in $\varphi(\bar{x})$) such that $\psi(\bar{y})$ is equivalent to $R_\Delta(\varphi(\bar{x}) \wedge \neg\, \delta(\bar{x},\bar{y})) < \alpha$. Thus $\mathbb{C} \models \psi(\bar{b})$ iff $\text{R-m}_\Delta(\varphi(\bar{x}) \wedge \delta(\bar{x},\bar{b})) \models (\alpha,m)$. □

Lemma 3.7. $R_\Delta(\varphi(\bar{x})) < \omega$ for all $\varphi(\bar{x})$.

Proof: If not there is $\varphi(\bar{x})$ with $\text{R-m}_\Delta(\varphi(\bar{x})) = (\omega,1)$. By fact 3.5 (ii) and the finiteness of Δ there is $\delta(\bar{x},\bar{y}) \in \Delta$ such that for arbitrarily large $r < \omega$ there is $\bar{b}$ such that $R_\Delta(\varphi(\bar{x}) \wedge \delta(\bar{x},\bar{b})) \geq r$, $R_\Delta(\varphi(\bar{x}) \wedge \neg\, \delta(\bar{x},\bar{b})) \geq r$. But, as in proof of 3.6, $R_\Delta(\varphi(\bar{x}) \wedge \delta(\bar{x},\bar{b})) \geq r$ is equivalent to $\Gamma_r(\bar{b})$ for some collection $\Gamma_r(\bar{x})$ of formulas, and similarly for "$R_\Delta(\varphi(\bar{x}) \wedge\neg\, \delta(\bar{x},\bar{b})) \geq r$". There is clearly $\bar{b}$ such that $R_\Delta(\varphi(\bar{x}) \wedge \delta(\bar{x},\bar{b})) \geq \omega$ and $R_\Delta(\varphi(\bar{x}) \wedge \neg\, \delta(\bar{x},\bar{b})) \geq \omega$, contradicting $\text{R-m}_\Delta(\varphi(\bar{x})) = (\omega,1)$. □

Let $p(\bar{x}) \in S(M)$. We say p is <u>definable</u> if for every $\varphi(\bar{x},\bar{y}) \in L$ there is $\psi(\bar{y})$ over M with : for $\bar{b} \subset M$, $\varphi(\bar{x},\bar{b}) \in p$ iff $M \models \psi(\bar{b})$. If the $\psi(\bar{y})$ are all over $A \subset M$ we say p is definable over A. Note this means that there is a defining schema over A, d say, such that $d(M) = p$.

Proposition 3.8. (i) Every $p(\bar{x}) \in S(M)$ is definable.

(ii) Let $p(\bar{x}) \in S(A)$ where A is algebraically closed (in $\mathbb{C}^{eq}$). Then for any finite $\Delta(\bar{x})$ there is a Δ-defining schema d over A such that $p(\bar{x}) \cup d(M)$ is consistent (M any model containing A).

Proof: (i) Let $p(\bar{x}) \in S(M)$. Let $\varphi(\bar{x},\bar{y}) \in L$. Let $\Delta(\bar{x}) = \{\varphi(\bar{x},\bar{y})\}$. Let $\chi(\bar{x}) \in p$ be such that $\text{R-m}_\Delta(\chi(\bar{x})) = (k,m)$ is least possible (in fact $m = 1$). So for $\bar{b} \in M$, $\varphi(\bar{x},\bar{b}) \in p$ iff $\text{R-m}_\Delta(\chi(\bar{x}) \wedge\varphi(\bar{x},\bar{b})) = (k,m)$ which, by

Lemma 3.6, is equivalent to $\models \psi(\bar{b})$ for a formula $\psi(\bar{y})$ with parameters among those in χ.

(ii) Let $p(\bar{x}) \in S(A)$. Let $M \supset A$ be saturated. By Fact 3.5 (iii) there is a <u>complete Δ-type</u> $q(\bar{x})$ over M with $R_\Delta(p(\bar{x})) = R_\Delta(p(\bar{x}) \cup q(\bar{x}))$. By the same proof as in (i) there is a Δ-defining schema over M, d say, such that $q = d(M)$. We will show that d is over A (i.e. q is definable over A). Let f be an automorphism of M which fixes A pointwise. So $f(p \cup q) = p \cup f(q)$ and $R_\Delta(p) = R_\Delta(p \cup f(q))$. So by definition of R_Δ, q has only finitely many images under such A-automorphisms of M. Thus for each $\delta(\bar{x},\bar{y}) \in \Delta$, the defining formula $d(\delta(\bar{x},\bar{y}))$ has only finitely many images under A-automorphisms of M, i.e. $d(\delta(\bar{x},\bar{y}))$ is almost over A. But A is algebraically closed in $\mathbb{C}^{eq}$ so $d(\delta(\bar{x},\bar{y}))$ is over A. This shows that d is a Δ-defining schema over A. □

Proposition 3.9. Let $p(\bar{x}) \in S(A)$, A algebraically closed. Let d_1, d_2 be Δ-defining schema over A such that for $M \supset A$, $p(\bar{x}) \cup d_1(M)$ is consistent and $p(\bar{x}) \cup d_2(M)$ is consistent. Then $d_1(M) = d_2(M)$ (all $M \supset A$).

Proof: Let $\varphi(\bar{x},\bar{y}) \in \Delta$. Let $\delta_1(\bar{y}) = d_1(\varphi(\bar{x},\bar{y}))$ and $\delta_2(\bar{y}) = d_2(\varphi(\bar{x},\bar{y}))$. We must show that $M \models (\forall \bar{y})(\delta_1(\bar{y}) \leftrightarrow \delta_2(\bar{y}))$. Let $\bar{b} \subset M$ with $\ell(\bar{b}) = \ell(\bar{y})$. Let $q(\bar{y}) = tp(\bar{b}/A)$. Let $\Delta'(y)$ be a finite set of L-formulas containing $\varphi(\bar{x},\bar{y})$. By Proposition 3.8 (ii) there is a Δ'-defining schema d_3 over A such that $q(\bar{y}) \cup d_3(M)$ is consistent. Let $\delta_3(\bar{x})$ be $d_3(\varphi(\bar{x},\bar{y}))$. By Lemma 3.2, $\delta_1(\bar{y}) \in q(\bar{y})$ iff $\delta_3(\bar{x}) \in p(\bar{x})$ iff $\delta_2(\bar{y}) \in q(\bar{y})$. In particular $\models \delta_1(\bar{b}) \leftrightarrow \delta_2(\bar{b})$, which proves what is required. □

Corollary 3.10. Let $p(\bar{x}) \in S(A)$ (A algebraically closed). Then there is a <u>unique</u> defining schema d over A such that for all $M \supset A$, $p(\bar{x}) \cup d(M)$ is consistent (i.e. $p(\bar{x}) = d(A)$).

Proof: By 3.8 (ii) and 3.9. □

Note: Corollary 3.10 says that for any $p \in S(A)$ (A algebraically closed) and $M \supset A$ there is a unique $q \in S(M)$ such that $q \supset p$ and q is definable over A.

Definition 3.11. For $p \in S(A)$ (A algebraically closed), we denote by d_p the unique defining schema over A given by 3.10. Let $q \in S(B)$ and $A \subset B$ (A, B not necessarily algebraically closed). We say q does not fork over A if for some (equivalently <u>any</u>) extension q_1 of over acl(A), d_{q_1} is over acl(A).

Remark 3.12. So $q \in S(B)$ does not fork over A iff for some (any) extension q_1 of q over acl(B), d_{q_1} and $d_{q_1 \restriction acl(A)}$ are equivalent. This is by the uniqueness part of 3.10. It is also the case that $q \in S(B)$ does not fork over A iff $R_\Delta(q) = R_\Delta(q \restriction A)$.

Properties (i) - (vi) mentioned in the introduction to this section are now more or less immediate. Note that property (iv) (symmetry) follows from Lemma 3.2. We should also add

Corollary 3.13. For any $p \in S(A)$ and $B \supset A$ there is $q \in S(B)$ $q \supset p$, q does not fork over A (q is called a nonforking extension of p).

And we summarize:

Corollay 3.14. (i) Let $p(\bar{x}) \in S(M)$ and $A \subset M$. p does not fork over A iff p is definable over acl(A).

(ii) Let $p(\bar{x}) \in S(A)$, A algebraically closed. For any $B \supset A$, p has a <u>unique</u> nonforking extension to B.

Proposition 3.15. Let $p(\bar{x}) \in S(M)$, $A \subset M$ and suppose p does not fork over A. Let $\varphi(\bar{x},\bar{b}) \in p(\bar{x})$ $(\bar{b} \subset M)$. Then for every model $M_1 \supset A$, there is $\bar{a}' \in M_1$, for which $\models \varphi(\bar{a}',\bar{b})$ (i.e. p is <u>almost</u> finitely satisfiable in M_1).

Proof: We may assume A to be algebraically closed (in $\mathbb{C}^{eq}$) (as $A \subset M_1$ iff $\mathrm{acl}\, A \subset M_1$). Now let $M_1 \supset A$ and let p' be a nonforking extension of p over $M \cup M_1$ (by Corollary 3.13). So clearly p' does not fork over M_1. Let $\bar{a}$ realize p' (so $\models \varphi(\bar{a},\bar{b})$). By symmetry (iv) $tp(\bar{b}/M_1 \cup \bar{a})$ does not fork over M_1. So $tp(\bar{b}/M_1 \cup \bar{a}) = d(M_1 \cup \bar{a})$ for some defining schema over M_1. Let $\psi(\bar{x})$ be $d(\varphi(\bar{x},\bar{y}))$ ($\psi(\bar{x})$ has parameters M_1). Note $\models \exists \bar{x}\, \psi(\bar{x})$ (namely $\bar{a}$). Thus there is $\bar{a}' \in M_1$ $\models \psi(\bar{a}')$, so $\models \varphi(\bar{a}',\bar{b})$. □

We say $p(\bar{x}) \in S(A)$ is <u>stationary</u> if p has a unique nonforking extension over any $B \supset A$.

Lemma 3.16. $p \in S(A)$ is stationary if and only if for some $M \supset A$ and some $q \in S(M)$ containing p, q is definable over A.

Proof: Let $M \supset A$ be saturated (homogeneous). Let $q(\bar{x}) \in S(M)$ be a nonforking extension of p. So we know $q = d(M)$, where d is a defining schema over acl(A). Now for any A-automorphism f of M, $f(q) = f(d)(M)$ still extends p, so is clearly also a nonforking extension of p. By hypothesis $f(q) = q$, i.e. $f(d) = d$. It clearly follows that d is over A.

Conversely, suppose the right hand side to be true. Let $q \in S(M)$, $q \supseteq p$, q definable over A. So $q = d(M)$ for some defining schema d over A. By the uniqueness in Corollary 3.10 $d = d_p$, which implies that p is stationary. □

Lemma 3.17. $tp(\bar{a}/A)$ is stationary iff $\mathrm{dcl}(A,\bar{a}) \cap \mathrm{acl}(A) = \mathrm{dcl}(A)$ (where acl, dcl are computed in $\mathbb{C}^{eq}$).

Proof: Assume $tp(\bar{a}/A)$ to be stationary. Let $c \in \mathrm{dcl}(A,\bar{a})$. Easily $tp(c/A)$ is stationary. Thus if $c \in \mathrm{acl}(A)$ then $c \in \mathrm{dcl}(A)$.

On the other hand, suppose the right hand side is satisfied. Let $M \supseteq A \cup \bar{a}$ be saturated and let $q(\bar{x}) \in S(M)$ be a nonforking extension of $tp(\bar{a}/\mathrm{acl}A)$. Let $q(\bar{x}) = d(M)$ where d is a defining schema over aclA. Let f be an $(A \cup \bar{a})$-automorphism of M. So f(aclA) = aclA (as a set), and so f(q) = q, so f(d) = d. Thus d is over $A \cup \bar{a}$ i.e. for every $\varphi(\bar{x},\bar{y}) \in L$, $d\varphi(\bar{x},\bar{y})) \in dcl(A \cup \bar{a}) \cap aclA$. By the condition, $d(\varphi(\bar{x},\bar{y})) \in dclA$, i.e. d is over A. By 3.10, p is stationary. □

We now examine what some of these notions mean in the context of ω-stable theories and algebraically closed fields.

Lemma 3.18. Let T be ω-stable. Then T is stable. If $p \in S(A)$, $p \subset q \in S(B)$, then q is a nonforking extension of p iff RM(q) = RM(p). p is stationary iff Morley degree of p is 1.

Proof: $RM(\varphi(\bar{x})) < \infty$ clearly implies $R_\Delta(\varphi(\bar{x})) < \infty$ for all finite Δ. So T is stable.

It is easy to see that $RM(tp(\bar{a}/A)) = RM(tp(\bar{a}/aclA))$. So we may assume A to be algebraically closed. Also we may assume B is a big model M say. Let $q \in S(M)$, RM(q) = RM(p). Any conjugate of q under an A-automorphism of M clearly has the same property, so there are only finitely many such q. Thus q is definable over aclA = A. Hence q does not fork over A. The converse is similar. Clearly p has Morley degree 1 iff p has a unique extension to M with the same Morley rank, which by the above means that p is stationary. □

In the next observation we will use the fact, pointed out by Poizat in [Po1] that algebraically closed fields admit so-called elimination of imaginaries. This means that if K is an algebraically closed field and $a \in K^{eq}$ then there is some k-tuple $(b_1,...,b_k)$ from K much that $\bar{b}$ and a are interdefinable (over ∅) in K^{eq}. We also use the fact that if K is an algebraically closed field of characteristic 0 and $A \subset K$, then the definable closure in K of A equals the subfield of K generated by A (i.e. the rational closure of A). (By quantifier elimination).

Corollary 3.19. (of Lemma 3.17). Let K be an algebraically closed field of characteristic 0, $\bar{a} \subset K$ and k a subfield of K. Then $tp(\bar{a}/k)$ is stationary iff k is algebraically closed in $k(\bar{a})$ (i.e. $k(\bar{a})$ is a regular extension of k).

Proof: By the above remarks $dcl(k \cup \bar{a})$ in K^{eq} is interdefinable with $k(\bar{a})$ (the rational closure of $k \cup \bar{a}$). Similarly, acl(k) in K^{eq} is interdefinable with $\bar{k}$ (the algebraic closure of k in K). Thus the condition in Lemma 3.17 translates into: k is algebraically closed in $k(\bar{a})$, proving the Corollary. □

Let now K be a saturated algebraically closed field (of arbitrary characteristic). Fix $n < \omega$. We have already remarked that the affine algebraic subsets of K^n are the subsets of K^n defined by $P_1(\bar{x}) = 0 \wedge ... \wedge P_k(\bar{x}) = 0$, where $P_i \in K[\bar{x}]$. We point out that these are the closed sets for a certain Noetherian topology on K^n, the Zariski topology. It is first easy to see that a finite union of affine algebraic sets is also an affine algebraic set. On the other hand, if $V_i \subseteq K^n$, $i < \omega$, are affine algebraic sets then the ideal of $K[\bar{x}]$ generated by all the polynominals defining all the sets V_i, is generated by finitely many such polynomials (as $K[\bar{x}]$ is a Noetherian ring) and thus the intersection of the V_i is a finite subintersection. This shows that we have the DCC on affine algebraic sets. So we call the affine algebraic subsets of K^n, the Zariski closed subsets of K^n and we see that this equips K^n with a Noetherian topology (the Zariski topology). A Zariski closed set V is said to be irreducible if we cannot write V as $V_1 \cup V_2$, where $V_i \subsetneq V$ $(i = 1,2)$ are also Zariski closed. It is standard to show that any Zariski closed $V \subseteq K^n$ can be written uniquely as a union of irreducible Zariski closed subsets.

Proposition 3.20. Let $k \subset K$, k perfect. Let $\bar{a} \in K^n$ and let $V = V(\bar{a}/k)$. Then the number of irreducible components of V = Morley degree of V = Morley degree of $tp(\bar{a}/k)$.

Proof: (Remark: We take k perfect, so that "defined over k" has the same meaning in model theory as in geometry). Let $p = tp(\bar{a}/k)$. Let $p_1,...,p_r$ be the nonforking extensions of p to K. (i.e. $p_i \in S_n(K)$, p_i does not fork over k). So r = Morley degree of p. For each i = 1,..., r let $V_i = V(p_i)$ = smallest Zariski closed set in p_i. Clearly V_i is irreducible (as K is a model) and $V_i \subseteq V$. By quantifier elimination p_i is "determined" by V_i. So if f is a k-automorphism of K then for any i, $f(V_i) = V_j$ for some j. Thus each V_i has only finitely many conjugates over k. Fix say V_1 and let $V_1 = V_{i_1}$, $V_{i_2},\ldots,V_{i_s}$, be the k-conjugates of V_1. Then $V_{i_1} \cup \ldots \cup V_{i_s}$ is defined over k, is in $p = tp(\bar{a}/k)$ and so equals V. By the irreducibility of each V_i this shows that $V_1 \cup \ldots \cup V_r = V$ and that the V_i are the irreducible components of V. This suffices to prove the proposition. □

Corollary 3.21. (Again k perfect). Let $V \subset K^n$ be an irreducible Zariski closed set, defined over k. Then there is $\bar{a} \in k^n$ with $V = V(\bar{a}/k)$. Also if $V_1 \subsetneq V$ is Zariski closed then $RM(V_1) < RM(V)$.

Proof: By irreducibility of V and compactness there is $\bar{a} \in K^n$ with $V = V(\bar{a}/k)$. Suppose $V_1 \subsetneq V$, $RM(V_1) = RM(V)$. Let p' be (by Prop 3.20) the nonforking extension of p to K. So, as Morley degree of V = 1, $V_1 \in p'$. But then $V_2 = V(p')$ is contained in V_1 and so is a proper irreducible component of V, and by the proof of 3.20 we obtain a contradiction. □

Corollary 3.22. (k perfect). Let $V \subseteq K^n$ be irreducible and defined over k. Let $\bar{k}$ be the algebraic closure of k (in K) and let $V(\bar{k})$ be $V \cap \bar{k}^n$. Then $V(\bar{k})$ in Zariski dense in V.

Proof: By 3.21 there is an $\bar{a} \in K^n$ with $V = V(\bar{a}/k)$. Let $p = tp(\bar{a}/k)$ and p' the nonforking extension of p to K. Let $X \subseteq V$ be Zariski open in V. So by Corollary 3.21, RM(X) = RM(V), so $X \in p'$ does not fork over k and $\bar{k}$ is an elementary substructure of K containing k. By Proposition 3.15 $X \cap \bar{k}^n$ ($= X \cap V(\bar{k})$) is nonempty. □

Let me finally in this section mention superstability a property stronger than stability but weaker than ω-stability. Again it can be defined by means of a rank. We define the rank R^∞ on definable subsets X of a very saturated model N, (the crucial clause being: $R^\infty(X) \geq \alpha + 1$ if there are for all λ, X_i $(i < \lambda)$, all defined by an instance of the same formula such that

(i) $R^\infty(X \wedge X_i) \geq \alpha$ and

(ii) the X_i are m-inconsistent for some $m < \omega$, i.e. for distinct $i_1,\ldots,i_m$ we have $X_{i_1} \wedge \ldots \wedge X_{i_m} = \emptyset$.

T is superstable if $R^\infty(X)$ is defined for all A. A rather different kind of rank, the U-rank (of Lascar) can be defined for complete types $p \in S(A)$ in a stable theory: $U(p) \geq \alpha + 1$ if p has a forking extension q such that $U(q) \geq \alpha$. It turns out that T is superstable if and only if $U(p) < \infty$ for all p.

Both ranks R^∞ and U reflect forking in a superstable theory: namely for $R = R^\infty$ or U and $p \subset q$ complete types, $R(p) = R(q)$ just if q is a nonforking extension of p.

4. ω-STABLE GROUPS

A stable group is a group $(G,\cdot,\ldots)$ equipped with possibly additional structure such that the theory of this structure is stable. Similarly for ω-stable groups, superstable groups. One could also (and we do) consider a stable group G as a group definable in a stable structure M; namely both the universe of the group G and the group operation are definable in M. This is the case of say affine algebraic groups over an algebraically closed fields K, such groups being definable (in K) subgroups of $GL_n(K)$. The two points of view amount to the same thing. For suppose G to be defined in the stable structure M; where $\varphi(\bar{x},\bar{a})$ defines the universe of G and $f(\bar{x},\bar{y},\bar{a})$ defines the group operation on G $(\bar{a} \subset M)$. For each relation on G^n defined in M by a formula with parameter $\bar{a}$, introduce a new relation symbol. Let $\underline{G}$ be G equipped with its multiplication and all there relations. By virtue of definablility of types in M, every definable in M subset of G^n is definable also in the structure $\underline{G}$. $\underline{G}$ inherits all the stability theoretic properties of M (stability, superstability, ω-stability and even $\aleph_1$-categority). Similarly for reducts of G (e.g. the pure group $(G,\cdot)$), except that $\aleph_1$-categority may no longer hold: as pointed out in a paper of Baldwin in this volume $GL_2(\mathbb{C})$ is not $\aleph_1$-categorical.

Examples of stable groups are: Abelian groups (as pure groups), modules, affine algebraic groups over algebraically closed fields (which are ω-stable of finite Morley rank), algebraic matrix groups over any stable ring, Abelian varieties (equipped with their induced structure from the underlying field).

We will specialize first to ω-stable groups, where the proofs of basic properities (generic types etc.) are somewhat easier, and then say a few words about general stable groups.

So let G be a group (with additional structure). At times we want to consider G as an elementary substructure of a saturated G_1, and sometimes G is itself taken to be saturated. The main new fact given to us by working with a group rather than an arbitrary structure is "homogeneity" - in the sense that for every $a, b \in G$ there is a definable bijection of G with itself taking a to b (e.g. right multiplication by $a^{-1}b$). It is clear that any definable bijection of a structure M also acts on the definable sets and preserves "everything" of a definable nature, in particular Morley rank, degree etc. So this is true in particular of left and right multiplication by elements of G. Note that if $X \subset G$ is defined by $\varphi(\bar{x},\bar{b})$ then $a\cdot X$ is defined by $\varphi(a^{-1}x,\bar{b})$. G also acts on the 1-types $p \in S_1(G)$: if $a \in G_1 \succ G$ realizes p and $b \in G$, then $bp = tp(ba/G)$ and $pb = tp(ab/G)$. Similarly $p^{-1} = tp(a^{-1}/G)$. So

$$RM(p) = RM(p^{-1}) = RM(bp) = RM(pb). \quad (*)$$

Proposition 4.1. Let G be ω-stable. Then G has the DCC on definable subgroups.

Proof: Let $H_1 \subsetneq H_2 < G$. (H_1, H_2 definable subgroups of G). If H_1 has infinite index in H_2 then clearly $RM(H_1) < RM(H_2)$. If H_1 has finite index in H_2 then $RM(H_1) = RM(H_2)$ but Morley degree $(H_1) <$ Morley degree

(H_2). As every definable subset of G has ordinal valued Morley rank and integer valued Morley degree, there cannot be an infinite descending chain of definable subgroups. □

Corollary 4.2. If G is ω-stable, then G has a smallest definable subgroup of finite index. We call this G°, the connected component of G. □

Note that G° is ∅-definable because (i) G° is definable and (ii) G° can be described without reference to any parameters.

Definition 4.3. Let G be ω-stable. $p \in S_1(G)$ is said to be a generic type of G if RM(p) = RM(G) (= RM("x = x") in Th(G)).

Remark 4.4. (G ω-stable)

(i) There are only finitely many generic $p \in S_1(G)$ (and there is at least one.

(ii) if p is generic so is p^{-1} (by *).

(iii) G acts (by left or right translation) on the generic types of $S_1(G)$ (by *).

In fact:

Lemma 4.5. G acts (by left or right translation) transitively and definably on the set of generics of $S_1(G)$.

Proof: First, let $p, q \in S_1(G)$ be both generic types, and let $a, b \in G_1 \succ G$ realize p, q respectively such that a and b are independent over G. Let $c = ab^{-1}$.

Claim: $RM(c/G \cup b) = RM(a/G \cup b)$.

This is because $\varphi(x,\bar{d}) \in tp(c/G \cup b)$ iff $\varphi(xb^{-1},\bar{d}) \in tp(a/G \cup b)$ and $\varphi(x,\bar{d}) \in tp(a/G \cup b)$ iff $\varphi(xb,\bar{d}) \in tp\,(c/G \cup b)$ and $\varphi(x,\bar{d})$, $\varphi(xb,\bar{d})$ have the same Morley rank.

Let $\alpha = RM(G)$. We know that $RM(a/G \cup b) = \alpha$. Thus

$RM(c/G \cup b) = \alpha$. So $RM(c/G) = \alpha$. By 3.18 c and b are independent over G and so by 3.15 $tp(c/G \cup b)$ is finitely satisfiable in G. Now let $\varphi(x) \in p = tp(a/G)$ have Morley rank α and degree 1. (So p is the unique type in $S_1(G)$ containing $\varphi(x)$ and with Morley rank α). We clearly have $\models \varphi(cb)$. So there is $c' \in G$ such that $\models \varphi(c'b)$. But as $tp(c'b/G)$ has Morley rank α, $tp(c'b/G) = p$. So $p = c'q$. This proves transitivity of the action.

Let $P = \{p_1,\ldots,p_n\}$ be the generic types of G. To say that G acts definably on P we mean (in this specal case) for each i,j there is a formula $\varphi_{ij}(x)$ with parameters in G such that $G \models \varphi_{ij}(a)$ iff $ap_i = p_j$. So fix such i,j. Clearly $RM\, p_i = RM\, p_j = \alpha$. Let $\varphi(x,\bar{a}) \in p_i$ with $RM(\varphi(x,\bar{a})) = \alpha$ and degree of $\varphi(x,\bar{a}) = 1$. Clearly $ap_i = p_j$ just if $\varphi(a^{-1}x,\bar{a}) \in p_j$. But p_j is definable (Prop 3.8), so there is $\psi(z,\bar{a},\bar{c})$ $(\bar{a},\bar{c} \subset G)$ such that for all $a \in G$. $\varphi(a^{-1}x, \bar{a}) \in p_j$ iff $G \models \psi(a,\bar{a},\bar{c})$. This is enough. □

Corollary 4.6. (G ω-stable). The Morley degree of G (= number of generic $p \in S_1(G)$) = index of G° in G.

Proof: Let $\{p_1,\ldots,p_n\}$ be the set of generics in $S_1(G)$. Let $K = \{a \in G : ap_i = p_i\ \forall i = 1,\ldots,n\}$. By Lemma 4.5, K is definable, and clearly has finite index in G and moreover

$$n \leq |G/K| \leq |G/G^\circ|.$$

On the other hand, each coset of G° in G has Morley rank α and so gives rise to at least one generic type, whereby $|G/G^\circ| \leq n$. Thus we have equality, proving the Corollary. □

Proposition 4.7. Let G be ω-stable with $RM(G) = \alpha$. Let $X \subseteq G$ be definable. Then $RM(X) = \alpha$ iff finitely many translates (left or right) of X cover G.

Proof: As finitely many translates of G° cover G, we may assume that $G = G^\circ$ (i.e. G is connected), and so by 4.6, G has a unique generic type. The right to left direction of the proposition is easy (as all translates of X have same Morley rank). For the left to right direction: Let $RM(X) = \alpha$. We will show that for all $q \in S_1(G)$ there is an element $c \in G$ such that the definable set cX is in q. Fix $q \in S_1(G)$ and a realization b of q. Let a realize the generic type of G such that a and b are independent over G. As in the proof of Lemma 4.5, $RM(ab/G) = \alpha$ ($= RM(G)$). As G has Morley degree 1 and $RM(X) = \alpha$, $ab \in X$. By independence and 3.15. $\exists\, a' \in G$ such that $a'b \in X$. i.e. $(a')^{-1} \cdot X \in q$. Put $c = (a')^{-1}$.

By compactness, finitely many translates of X cover G. □

So we can define a generic formula (or definable set) in G to be one finitely many left translates of which cover G.

Corollary 4.8. (G ω-stable) $p \in S_1(G)$ is generic iff p contains only generic formulas.

For stable groups G, Corollary 4.8 can be taken as a definition of generic type.

Proposition 4.9. Let G be ω-stable and connected. Let $X \subset G$ be generic. Then $X \cdot X = G$.

Proof: Let $a \in G$. Let $b \in G_1 > G$ realize the generic type of G. Let $\varphi(x)$ be the formula defining X. Now $tp(b^{-1}/G)$ and so also $tp(b^{-1}a/G)$ are generic. Thus $\models \varphi(b^{-1}a)$ and of course $\models \varphi(b)$. Thus $G_1 \models \exists x \exists y (\varphi(x) \wedge \varphi(y) \wedge a = xy)$. The same is true in G (as $a \in G$). So $X \cdot X = G$. □

Borel proved a useful fact about algebraic groups (over algebraically closed fields) which is the following: Let G be an algebraic group. Let X_i $i \in I$ be a family of constructible subsets of G, each containing the identity element e such that the Zariski closure $\overline{X}_i$ of each X_i is irreducible. Then the subgroup H of G generated by the X_i is (Zariski)-closed (i.e. constructible), connected and $H = X_{i_1}^{\epsilon_1} \dots X_{i_n}^{\epsilon_n}$ fo some $i_1,\dots,i_n \in I$, where $\epsilon_j = \pm 1$.

The proof is so direct that it is worth giving: One first observes that for all $i_1,\ldots,i_k \in I$ and $\epsilon_1,\ldots,\epsilon_k = \pm 1$ the Zariski closure of $X_{i_1}^{\epsilon_1},\ldots,X_{i_k}^{\epsilon_k}$ is irreducible. Thus there is $X = X_{i_1}^{\epsilon_1}\ldots X_{i_k}^{\epsilon_k}$ such that $\overline{X}$ is greatest. Easily $\overline{X}$ is a subgroup of G. As $\dim X = \dim \overline{X}$ and $\overline{X}$ is connected, by Proposition 2.9 even $X \cdot X = \overline{X}$.

Zil'ber remarkably proved a generalization of this result to ω-stable groups of finite Morley rank. Hrushovski in a paper in this volume proves the result in an even more general context. Here we give Zil'ber's proof. The problem of course is that in the general situation of ω-stable groups we have no geometry (at least a priori), so no notion of irreducible. Zil'ber finds a substitute for this: he calls definable $X \subset G$ indecomposable if for any definable subgroup H of G, either $|X/H| = 1$ or $|X/H|$ is infinite.

Proposition 4.10. Let G be ω-stable with finite Morley rank. For $i \in I$ let X_i be an indecomposable definable subset of G containing the identity element e. Let H be the subgroup of G generated by the subsets X_i. Then H is definable, connected and is equal to $X_{i_1}\ldots X_{i_k}$ for some $i_1,\ldots,i_k \in I$.

Proof: As RM(G) is finite, we can choose $i_1,\ldots,i_k \in I$ such that $X = X_{i_1}\ldots X_{i_k}$ has maximum possible Morley rank, say m. Let $p \in S_1(G)$, $X \in p$, $RM(p) = m$. Let $H = Fix(p)$ ($= \{a \in G : ap = p\}$) which is definable as in the proof of 4.5 (H is clearly a subgroup of G).

Claim (i): $X_i \subset H$ for all $i \in I$ (so H contains the subgroup generated by the X_i).

Proof: Fix $i \in I$. as $e \in X_i \cap H$ and X_i is indecomposable if $X_i \not\subset H$ then there would be $a_j \in X_i$ for $j < \omega$ such that $a_j \neq a_{j'} \bmod H$ for $j \neq j'$. As $H = Fix\ p$, it follows that $a_jp \neq a_{j'}p$ for $j \neq j'$. As $X_iX \in a_jp$ for all $j < \omega$, it follows that $RM(X_iX) \geq m + 1$, contradicting the choice of X. Thus Claim (i) is established.

Claim (ii): H is connected and p is the generic type of H.

Proof: Note that by Claim (i) $X \subset H$ and thus "$x \in H$" $\in$ p. As in the proof of 4.7, we can find $c \in H$ such that cp is a generic of H. By definition of H (as Fix(p)) we see that p is already a generic of H. By Lemma 4.5, it is the unique generic of H and thus by 4.6, H is connected. Thus we see that RM(H) = m, RM(X) = m and H is connected. By 4.9. H = X·X. □

Macintyre [M] proved that an ω-stable field is algebraically closed. The model theoretic ingredients of this are:

Lemma 4.11: Let K be an ω-stable infinite field. Then K has Morley degree 1 (so by 4.6 is connected both additively and multiplicatively).

Proof: Suppose A is a proper additive subgroup of K of finite index. So $\bigcap \{kA : k \in K\}$ is an <u>ideal</u> I of K. But by 4.1 $I = k_1A \cap ... \cap k_nA$ for some $k_1,...,k_n \in k$, so I has finite index in K. Since K does not have nontrivial ideals, I = K. So A = K. Thus by 4.6, K has Morley degree 1.

□

Lemma 4.12. Let G be a connected ω-stable group and f: $G \rightarrow G$ a definable endomorphism with finite kernel. Then f is surjective.

Proof: G/Kerf is definably isomorphic to lmf, the latter being a definable subgroup of G. As Kerf is finite, properties of Morley rank imply that RM(G) = RM(lmf). By connectedness of G, lmf = G. □

Now, by considering the maps $x \rightarrow x^n$ (endomorphism of K* with finite kernel) and, if charK = p, the maps $x \rightarrow x^p - x$ (endomorphism of K with finite kernel) we see from the previous two lemmas that K is perfect,

$x^n - a = 0$ has a root in K $\forall a \in K, \forall n < \omega$, and if char K = p, $x^p - x - a = 0$ has a root in K $\forall a \in K$. Moreover this is also true for every finite extension of K (as any finite extension of K is interpretable in K and thus also ω-stable). Now Galois theory implies that K is algebraically closed.

One of the important applications of Zil'ber's indecomposability theorem is to find a field in certain algebraic situations of finite Morley rank. The following proposition summarizes the essential points.

Before stating and proving this we make a few explanatory remarks: First let G be an ω-stable group, $A \subset G$ and $a \in G$; we say that a is generic over A if tp(a/A) has a nonforking extension $p \in S_1(G)$ which is generic (Note that if $p \in S_1(G)$ is generic then p does not fork over $\emptyset$ so $\forall A \subset G$ $p \restriction A$ is "generic").

Secondly let G, A be definable groups in an ω-stable structure such that G acts definably on A, as a group of automorphisms of A (e.g. A is a normal subgroup of G and G acts by conjugation). We call $X \subset A$ G-invariant if X is fixed setwise by G.

Fact 4.13: Let $X \subset A$ be G-invariant. Then X is indecomposable if and only if for every definable G-invariant subgroup H of A we have $|X/H| = 1$ or ∞.

Proof: Let H be an arbitrary definable subgroup of A. By the DCC $\bigcap_{s \in G} H^s = H^{s_1} \cap ... \cap H^{s_n}$ for some $s_1,...,s_n \in G$. If X is G-invariant and $|X/H| < \omega$ then clearly $|X/(H^{s_1} \cap ... \cap H^{s_n})| < \omega$ and note that $H^{s_1} \cap ... \cap H^{s_n}$ is G-invariant. □

Proposition 4.14. Let G, A be (infinite) definable Abelian groups in a structure M of finite Morley rank and assume that G acts definably and faithfully on A. Suppose moreover that A has no infinite proper G-invariant subgroup. Then there is in M a definable field R such that the additive group of R is definably isomorphic to A, G definably embeds into the multiplicative group of R, and the action of G on A corresponds to multiplication in R.

Proof: We first note that A must be connected. Now let a be a generic of A over $\emptyset$. We claim that $G \cdot a$ is infinite. For otherwise $G^\circ \cdot a$ is finite, so $G^\circ \cdot a = \{a\}$ (as G° is connected), but then as every element of A is a sum of generics, $G^\circ b = b \ \forall b \in A$ so $G^\circ = \{1\}$, contradicting G being infinite.Now $G \cdot a$ is clearly G-invariant. By Fact 4.13, $Ga \cup \{0\}$ is an indecomposable subset of A. By Proposition 4.10, there is an integer $n < \omega$ such that for every $b \in A \ \exists g_1,\ldots,g_n \in G$ such that $b = g_1 \cdot a + \ldots + g_n \cdot a$ (**).
Let R be the subring of End(A) generated by G. So R is commutative and by (**) every element r of R is determined by its action on a; in fact every $r \in R$ is of the form $g_1 + \ldots + g_n$ for some $g_1,\ldots,g_n \in G$. It easily follows that R and its action on A are definable, using a as a parameter. We claim that R is a field. We must show that every nonzero element of R has an inverse in R. Let $r \in R$ $r \neq 0$. Now Ker r is a G-invariant subgroup of A, so must be finite. By Morley rank considerations r is surjective. Let therefore $b \in A$ be such that $r \cdot b = a$ where a is the generic element of A chosen above. By (**) there is $s \in R$ such that $sa = b$. So $rs(a) = a$. By (**) again $rs = 1$. This shows that R is a field. Clearly $G \subset R$, and the map $r \rightarrow ra$ is an additive isomorphism between R and A.

□

Corollary 4.15. Let G be a connected ω-stable group of finite Morley rank which is solvable but non nilpotent. Then G interprets an infinite field.

Proof: First a remark: If G is connected and Z(G) is finite then G/Z(G) is centreless. For if $a \in G$ is such that a is central mod Z(G) then $a^{-1} \cdot a^G \subset Z(G)$, so a^G is finite, so $C_G(a)$ has finite index in G. Thus $C_G(a) = G$ and $a \in Z(G)$.

Now suppose G to be solvable, non nilpotent and connected of finite Morley rank. If Z_n is the upper central series, then since G has finite Morley rank, Z_{n+1}/Z_n is finite for some n. But then G/Z_{n+1} is centerless by the above remark. Clearly $G \neq Z_{n+1}$ (as G is nonnilpotent). So working now with G/Z_{n+1} (which remains solvable) we may assume G to be centreless (and still connected). Let A be a minimal infinite definable normal subgroup of G. Now A is connected and solvable, and by Prop 4.10 the derived group A' is definable and connected. Thus $A' = \{e\}$, A is abelian. As G has no center, $C_G(A) \neq G$. So $G/C_G(A)$ is infinite and acts faithfully on A. Let H be an infinite abelian definable subgroup of $G/C_G(A)$ (by ω-stability [Ch]) and let B be a minimal infinite H-invariant subgroup of A. H and B satisfy the conditions of Prop 4.14, thus an infinite field is interpreted. □

Before the next application of 4.14, we need to know something about automorphisms of fields of finite Morley rank.

Fact 4.16. Let K be a (infinite) field of finite Morley rank. Then K has no infinite definable proper subfield.

Proof: Let L be a definable infinite subfield. So L is ω-stable and thus algebraically closed. K could not therefore be a finite extension of L. But then L^n can be definably embedded in K for all $n < \omega$, whereby easily K must have infinite Morley rank. □

Lemma 4.17. Let K be a (infinite) field of finite Morley rank. Let α be a nontrivial definable field automorphism of K. Then

(i) α has infinite order

(ii) α is $acl(\emptyset)$-definable.

Proof: (i) If α had finite order then the fixed field of α would be infinite and definable, contradicting the previous fact.

(ii) If α were not $acl(\emptyset)$-definable then we could find β, a definable field automorphism, with $\alpha \neq \beta$ and $stp(\alpha) = stp(\beta)$. But then α and β must agree on the algebraic closure of the prime subfield of K, which is infinite. So the fixed field of $\beta^{-1}\alpha$ is infinite, but $\beta^{-1}\alpha \neq \mathrm{id}$, again contradicting the previous fact.

Corollary 4.18. A field K of finite Morley rank cannot have a definable group G of automorphisms.

The second major application of 4.14 is Nesin's theorem:

Proposition 4.19. Let G be solvable, connected of finite Morley rank. Then G' (the commutator subgroup) is nilpotent.

Proof: We suppose not and obtain a contradiction. As in the proof of 4.15, we may assume G' to be centreless. (Note G' is definable, connected by 4.10). Let A_1 be a minimal infinite normal (in G') definable subgroup of G'. By induction we can choose A_1 in the center of $(G')'$. As G' is solvable, A_1 is abelian. Let A be the subgroup generated by the A_1^g, $g \in G$. Then $A < G'$ is abelian, normal in G and in fact $A = A_1 \oplus A_1^{g_1} \oplus \ldots \oplus A_1^{g_k}$ some $g_1,\ldots,g_k \in G$ (using minimality of A_1, the fact that G' is centreless and finiteness of Morley rank). The proof now follows a series of steps: Let R be the ring of endomorphisms of A generated by G' (acting on A by conjugation) and let I be the ideal in R consisting of those $r \in R$ which annihilate A_1.

(1) R/I is an infinite field K, which is precisely the ring of endomorphisms of A_1 generated by $G'/C_{G'}(A_1)$. The latter, as well as its action on A_1 is definable, by 4.14. (Note $C_{G'}(A_1) \neq G'$ as G' is centreless). Similarly $I^g = \{r \in R : r(A_1^g) = 0\}$ is a maximal ideal of R for all $g \in G$.

(2) There are only finitely many ideals I^g of R for $g \in G$.

Proof: Suppose not; let $m>RM(A)$, and suppose $I^{g_1},\ldots,I^{g_m}$ are distinct ideals. Let $B_i = A_1^{g_1}$, let $b_i \in B_i$ such that $b_1 + \ldots + b_m = 0$. Let $r \in R$ be such that $r \equiv 1 \bmod I^{g_1}, r \in I^{g_i}$ for $i = 2,\ldots,m$ (as the I^g are maximal ideals). So $r(b_1+...+b_m) = rb_1 = 0$, so $b_1= 0$. Similarly $b_i= 0$ $\forall i$. Thus the subgroups B_i direct sum which implies that $RM(A) \geq m$, contradiction.

(3) $I = I^g$ for all $g \in G$.

Proof: Let by (2) $I_1, I_2, ..., I_n$ be the distinct conjugates of I. For all $m < \omega$ let R_m be those members of R that can be expressed as $h_1 + ... + h_m$ for $h_i \in G'$. Thus there is an m such that the $I_j \cap R_m$ are pairwise distinct. But G acts transitively and definably on the $I_j \cap R_m$. As G is connected, there is only one. So (3) is proved.

(4) $R = K$ and is action on A (making A into a K-vector space) is definable.

Proof: As A is generated by the A^g and $I = I^g$ $\forall g$, it follows that $I = 0$. Thus "R = K" and the action of $r \in R$ on A is determined by its action on A_1. More precisely: given that $I = 0$ it follows that the action of an element $r \in R$ on A is determined by its action on A_1. But the action of R on A_1 is precisely that of K. Now K is definable: every element k of K can be represented by $h_1 + ... + h_n$, for $h_i \in G'$ and fixed $n \in \omega$ (by 4.14). Multiplication and addition of such elements are definable using a parameter from A, as is the relation $h_1 + ... + h_n = h'_1 + ... + h'_n$. The action of an element of R represented by such $h_1 + ... + h_n$ on A is precisely $(h_1+...+h_n) \cdot a = h_1 \cdot a + ... + h_n \cdot a$. We now identify K with R.

It is easily checked that G acts as a group of automorphisms of K by $(h_1 + ... + h_n)^g$ and thus by Corollary 4.18 we have

(5) for every $k \in K$, $g \in G$ $k^g = k$.

Finally

(6) The action of G on A (by conjugation) is K-linear.

Proof: Let $k \in K$, $a \in A$, $g \in G$. Then clearly $(k\cdot(a))^g = k^g\cdot a^g = k\cdot a^g$, using (5).

By (6), G' acts on the K-vector space A as matrices with determinant 1. But G' also acts as scalar multiplication. Thus G' acts trivially on A i.e. $A \subset Z(G')$ which contradicts everything. □

A result whose proof uses similar ideas to the above is:

Proposition 4.20. Let G be solvable centreless connected of finite Morley rank. Let A be the socle of G (= group generated by minimal normal definable subgroups). Then the ring R of endomorphisms of A generated by the action of G on A by conjugation is definable, as well as the <u>action</u> of R on A.

Proof: As G' is nilpotent (by 4.19) it is clear that every minimal normal definable subgroup of G is in $Z(G')$. Moreover $A = A_1 \oplus A_2 \oplus ... \oplus A_n$ where each A_i is minimal normal definable. Then R is generated by $G/C_G(A)$, which is Abelian (as $C_G(A) \supset G'$), and so R is commutative. Let I_i = the ideal of R which annihilates A_i. As in 4.19, R/I_i is a field and is "identical" to the ring of endomorphisms of A_i generated by $G/C_G(A_i)$, which is definable by 4.15. Rewrite A as $B_1 \oplus ... \oplus B_k$ where $B_j = A_{j1} \oplus ... \oplus A_{jm_j}$, the ideals corresponding to the A_{ji} are the same and for $j_i \neq j_2$ the ideals corresponding to the $A_{j_1 i}$, $A_{j_2 i}$ are different. Rewrite I_j as the annihilator of A_j (=annihilator of B_j). As in the proof of 4.19, B_j is a R/I_j vector space, where R/I_j and its action on B_j are definable. As the I_j are maximal ideals of R, for each j there is $r_j \in R$ which is 1 mod I_j and 0 mod I_k for $k \neq j$. Writing $K_j = R/Ij$, it follows that for any $s_1 \in K_1, ..., s_k \in K_k$ there is $r \in R$ such that for $a = b_1 + ... + b_k \in A$, $r\cdot a = s_1\cdot b_1 + ... + s_k\cdot b_k$. Thus the action of R on A is the product of the actions of the K_j on B_j, and so is definable. □

The greatest and in some sense correct level of generality of stable group theory is of course stable groups. In place of the DCC on definable subgroups, one has the weaker DCC on intersections of uniformly definable subgroups.

Proposition 4.21. Let G be a stable group, and $\varphi(x,\bar{y})$ a formula. Let H_i $(i \in I)$ be subgroups of G, each defined by an instance of φ. Then $\bigcap_{i \in I} H_i = H_{i_1} \cap \ldots \cap H_{i_n}$ for some $i_1,\ldots,i_n \in I$.

Proof: Let $\Delta(x)$ a a finite collection of formulas including $\varphi(z \cdot x,\bar{y})$. If by way of contradiction, we had an infinite descending chain of finite intersections of the H_i; say $K_1 \supsetneq K_2 \supsetneq K_3 \ldots$, then for each i we would clearly have $R_\Delta(K_{i+1}) < R_\Delta(K_i)$ or $m_\Delta(K_{i+1}) < m_\Delta(K_i)$, which would be a contradiction to 3.2. □

Generic types in the general stable context can be defined using the notion of generic formula, introduced earlier: roughly p is generic iff it only contains generic formulas. This is equivalent to : if G is $|T|^+$-saturated and $p \in S_1(G)$, then p is a generic type of G if and only if ap does not fork over $\emptyset$ for all $a \in G$. A detailed exposition of the theory of generic types in the general stable situation appears in Victor Harnik's paper in this volume, so we will not go into any further details.

If G is stable and very saturated, we will call a subgroup H of G infinitely definable if H is defined by a collection of at most $|T|$ formulas. A theorem of Poizat asserts that the formulas can be taken to define subgroups of G. On the other hand, if H is the intersection of some arbitrary number of definable subgroups of G, then by 4.21, H is actually the intersection of at most $|T|$ many definable subgroups. In general infinitely definable subgroups arise in the stable context where definable subgroups appear in the ω-stable context. For example, if G is saturated, $p \in S_1(G)$, then Fix p is an infinitely definable subgroup of G, by virtue of definability of types.

Superstable groups are of course in between. On the one hand by looking at the rank R^{∞} mentioned at the end of Section 3, we see that superstable groups have no infinite descending chains of definable subgroups, each of infinite index in the proceding one. On the other hand, a superstable group may not contain a smallest definable subgroup of finite index. So infinitely definable subgroups only come into the picture when we want to obtain connected components: If G is a $|T|^+$–saturated superstable group then the connected component G° of G is the intersection of all definable subgroups of G of finite index. $R^{\infty}(G^{\circ}) = R^{\infty}(G)$. An important fact about superstable groups G is that the U-rank of G is well defined. There are types $p \in S_1(G)$ of maximum U–rank; these are precisely the generic types of G. An important topic that we have not mentioned is the study of groups of finite Morley rank from the point of view that they should resemble algebraic groups over algebraically closed fields. As this is clearly false for abelian groups (consider $Z_{p^{\infty}}$), some hypothesis of non-abelianness should be imposed. The major work was done by Cherlin [Ch], where he showed:

(1) A Morley rank 1 group is abelian-by-finite (Actually this is due to Reineke; Cherlin noted that any (infinite) ω-stable group has an infinite definable abelian subgroup).

(2) If G is connected, of Morley rank 2, then G is solvable. If also G is centreless then G is the semidirect product of the additive and multiplicative groups of an algebraically closed field.

(3) If G is connected of Morley rank 3 and G is nonsolvable centreless and G has a definable subgroup of Morley rank 2, then $G = PSL_2(K)$ for K an algebraically closed field.

The attempt to eliminate (i.e. prove) the condition that G has definable subgroup of rank 2 has led to an important line of research ([N1], [B.P], [Co]).

Notes for section 4. 4.1 is due to Macintyre [M] 4.6 is due to Cherlin [Ch] and Zilber [Z]. Our treatment of the material is influenced by Poizat's book [Po2]. Prop. 4.10 is due to Zilber [Z] as are 4.14, 4.15. Proposition 4.19, 4.20 are due to Nesin [N2]. 4.21 is Baldwin-Saxl [B.S]. The theory of stable groups in its full generality is due to Poizat.

References

[B.P] A.B. Borovik & B. Poizat, Mauvais Groupes, in Russian, submitted to Sib. Math. J.

[Ch] G. Cherlin, Groups of Small Morley Rank, Annals of Math. Logic 17 (1979), 1-28.

[Co] Corredor, Thesis, Bonn., see also Bad Groups of Finite Morley Rank, preprint.

[H.H] Harnik & L. Harrington, Fundamentals of Forking, Annals of Pure and Applied Logic 26 (1984) 245-286.

[L.P] D. Lascar, B. Poizat, Introduction to Forking, JSL, vol. 44, no. 3, Sept. 1979.

[M] A. Macintyre, ω_1-Categorical Fields, Fundamenta Mathematicae 70, no. 3 (1971), 253-270.

[N1] Ali Nesin, Non-Solvable Groups of Morley Rank 3, to appear in J. of Algebra (1988 +).

[N2] Ali Nesin, Solvable Groups of Finite Morley Rank, to appear in J. of Algebra (1988 +).

[P] A. Pillay, Introduction to Stability Theory. Oxford Univ. Press 1983.

[Po1] B. Poizat, Une Théorie de Galois Imaginaire, JSL 48, 1983, pp-1151-1170.

[Po2] B. Poizat, Groupes Stables, Nur Al-Mantiq Wal-Ma'rifah, Paris (1987).

[R] P. Rothmaler, Another Treatment of Forking Theory, Proceedings of 1st Easter Conference on Model Theory, Humboldt University, 1983.

[R-S] J. Baldwin and J. Saxl, Logical Stability in group theory, J. Australian Math. Soc. 21, 267-276.

[S] S. Shelah, Classification Theory and the number of Non-Isomorphic Models North Holland 1978.

[Z] B. Zil'ber, Groups and Rings with Categorical Theories, Fund. Math. 95 (1977) 173-188 (in Russian).

AN INTRODUCTION TO ALGEBRAICALLY CLOSED FIELDS & VARIETIES

Bruno Poizat
Université Pierre et Marie Curie

This introductory chapter has two aims: to explain to geometers what is the Logic point of view on algebraically closed fields, and to unveil to logicians the fundaments of algebraic geometry. It is expected that geometers and logicians will find here a common language; since Logic is far more direct than Geometry in the introduction of its basic concepts, this language will have a definite logical flavour; I hope that it will help the geometers to understand what arose the interest of logicians in algebraic groups, at an admitedly elementary level, but which nevertheless may bring some new contributions in this field.

I wrote this note when I was an invited professor in the University of Notre Dame du Lac; on this occasion, I witnessed how serious a matter was language tests for post-graduate students in the U.S.; this is the reason why I also designed this chapter to be an introduction to mathematical French, this beautiful language being so necessary for anybody who wish to have a deep insight either in Geometry or in Logic.

Corps algébriquement clos

Un corps algébriquement clos est un corps où les équations polynomiales ont autant de solutions qu'il est raisonable d'en espérer; K est algébriquement clos si et seulement s'il satisfait la conclusion du Théorème des Zéros de Hilbert: tout système formé d'un nombre fini d'équations et d'inéquations polynomiales, en plusieurs variables, à coefficients dans K, et qui a une solution dans une extension de K, a déjà une solution dans K.

La première chose que vous dira un logicien, à propos de cet énoncé, c'est qu'il fait intervenir une quantification portant sur toutes les extensions possibles de K; une telle quantification est dite "d'ordre supérieur". Pour de très bonnes raisons, que je ne vais pas justifier ici, les logiciens, lorsqu'ils décrivent une structure comme celle d'un corps K, aiment s'en tenir à des énoncés qui ne mentionnent que son addition +, sa multiplication ·, son zéro 0 et son unité 1, et qui ne comportent que des quantifications, dites du premier ordre, portant sur les éléments de K.

Et vous savez certainement que les corps algébriquement clos ont une définition équivalente plus simple, qui entre dans ce cadre: tout polynôme non-constant, en une variable, à coefficients dans K, a un zéro dans K. Strictement parlant, cet énoncé n'est du premier ordre qu'au prix d'une certaine hypocrisie, puisque le quanteur 'tout polynôme' porte non pas sur les éléments de K, mais sur les suites finies (de longueur arbitraire) d'éléments de K; on le remplacera donc par une collection infinie d'énoncés, un par degré possible du polynôme, qui ont sans conteste la forme requise:

$$(\forall a_0) \dots (\forall a_{n-1}) \ (\exists x) \ \ x^n + a_{n-1}x^{n-1} + \dots + a_0 = 0.$$

On voit d'ailleurs sans peine qu'un nombre fini d'entre eux ne peut suffire à impliquer la totalité des autres. Dans le même cadre entre l'expression de la caractéristique, et même par un énoncé sans quanteurs $1 + ... + 1 = 0$ si elle vaut p; pour la caractéristique nulle, il faut une infinité d'énoncés, tous ceux de la forme $1 + ... + 1 \neq 0$.

C'est un fait que deux corps algébriquement clos de même caractéristique satisfont les mêmes énoncés du premier ordre; c'est là l'expression rigoureuse d'un sentiment familier aux mathématiciens du début de ce siècle: si vous montrez une propriété des nombres complexes, par exemple en suivant une méthode analytique, et si cette propriété s' exprime par un énoncé du premier ordre ne parlant que de la somme et du produit de ce corps, alors elle est aussi vraie pour tout corps algébriquement clos de caractéristique nulle, et même pour tout corps algébriquement clos de caractéristique finie assez grande! La seule propriété d'un corps algébriquement clos qui échappe au premier ordre, c'est au fond ce qui le caractérise à isomorphie près, c'est-à-dire son degré de transcendance sur son corps premier.

Ensembles constructibles

Si je n'entreprends pas de justifier ce que j'ai dit au paragraphe précédent, ce n'est pas tellement pour éviter de nous égarer dans des considérations logiques, au demeurant peu difficiles, que parce que la Géométrie ne consiste pas à comparer des corps, mais à étudier les objets qu'on peut définir dans un corps K donné.

Les parties définissables les plus simples de K^n sont formées des zéros d'un certain polynôme $P(\overline{x})$ en n variables $\overline{x} = (x_1,...,x_n)$. L'équation $P(\overline{x}) = 0$ est considérée par les logiciens comme une "formule" définissant l'ensemble A de ses solutions; on peut insister sur les paramètres de cette équation, qui sont en l'occurence les coefficients du polynôme en question, en les faisant apparaitre dans l'écriture: $P(\overline{x},\overline{a}) = 0$; $\overline{x}$ et $\overline{a}$ ne sont pas vus de la même façon: le premier est un uple de variables, le second un uple de constantes, intervenant dans la formule servant à définir A, qui est l'ensemble des $\overline{x}$ annullant $P(\overline{x},\overline{a})$; évidemment rien n'interdit de considérer l'ensemble des $\overline{x} \wedge \overline{y}$ tels que $P(\overline{x},\overline{y}) = 0$!

Si A est défini par l'équation $P(\overline{x}) = 0$ et B par l'équation $Q(\overline{x}) = 0$, leur réunion $A \cup B$ est définie par la formule $P(\overline{x}) = 0 \vee Q(\overline{x}) = 0$, où $\vee$ désigne le symbole de disjonction. Dans ce cas particulier, on peut en faire l'économie, puisque cette formule équivaut à l'équation $P(\overline{x})\cdot Q(\overline{x}) = 0$. Par contre, on ne peut esquiver les symboles de conjonction quand on forme des intersections, et on appelle <u>fermé de Zariski</u> une partie A de K^n qui est définie par la conjonction d'un nombre fini d'équations: $P_1(\overline{x}) = 0 \wedge \ldots \wedge P_k(\overline{x}) = 0$.

En fait, par noethérianité de l'anneau des polynômes en n variables, un système formé d'une infinité d'équations peut se remplacer par un sous-système fini, si bien qu'une intersection quelconque de fermés de Zariski en est encore un; on voit aussi sans peine que la réunion de deux fermés de Zariski l'est encore. Mais le passage au complémentaire nous fournit de nouveaux ensembles, les <u>ouverts de Zariski</u>, définis par la disjonction d'un nombre fini d'inéquations: $P_1(\overline{x}) \neq 0 \vee \ldots \vee P_k(\overline{x}) \neq 0$. Si un ouvert de Zariski peut être défini par une seule inéquation on dit qu'il est <u>principal</u>.

Les géomètres qualifient de <u>constructible</u> un ensemble qui est combinaison booléenne finie de fermés de Zariski: il s'obtient à partir de ces derniers par applications successives de réunion, intersection et passage au complément; il est défini par une formule booléenne à partir d'équations polynomiales, et on voit facilement qu'on peut la mettre sous la forme d'une disjonction de systèmes formés d'un nombre fini d'équations et d'un seule inéquation:

$$\bigvee \ (P_1(\overline{x}) = 0 \ \wedge \ldots \wedge \ P_k(\overline{x}) = 0 \wedge \ Q(\overline{x}) \neq 0).$$

On a coutume de noter $\varphi(\overline{x})$ une telle formule, définissant un ensemble A de n-uples, ou $\varphi(\overline{x}\,;\,\overline{a})$ si on veut insister sur les paramètres, ou coefficients, qui y figurent. Si tous ces $\overline{a}$ sont dans le corps premier, engendré par 1, on dit que la formule est sans paramètres.

Ces ensembles constructibles sont qualifiés de <u>définissables</u> par les logiciens; d'habitude, ils qualifient ainsi les parties d'une structure définies par une formule où peuvent intervenir des quantifications du premier ordre. Si les quantifications ne sont pas apparues dans nos ensembles constructibles, c'est que, dans le cas particulier des corps algébriquement clos, il se trouve que tout ce qui se définit avec des quanteurs peut se définir sans. Ce phénomène est appelé élimination des quanteurs. Géomètres et logiciens le placent sous le vocable de deux saints patrons différents:

THEOREME DE CHEVALLEY: La projection d'un ensemble constructible est un ensemble constructible. (Note: si $A \subset K^n$ est défini par la formule $\varphi(x_1,...,x_{n-1},x_n)$, sa projection $B \subset K^{n-1}$ parallèlement à la dernière coordonnée est définie par $(\exists x_n)\ \varphi(x_1,...,x_{n-1},x_n)$).

THEOREME DE TARSKI: A toute formule $\varphi(\overline{x})$ du langage des corps est associée une formule $\psi(\overline{x})$ sans quanteurs de ce même langage telle que dans tout corps algébriquement clos K, $\varphi(\overline{x})$ et $\psi(\overline{x})$ définissent le même ensemble.

Ce résultat n'est en fait que l'avatar contemporain du plus ancien algorithme mathématique, connu depuis les temps babyloniens qui consiste à résoudre les systèmes d'équations et d'inéquations par élimination successive des inconnues, s'appuyant sur la remarque que la condition d'existence de l'inconnue éliminée s'exprime par la satisfaction de sytèmes d'équations et d'inéquations pour les inconnues restantes.

Pourquoi donc avoir empoisonné l'existence d'un lecteur géomètre par l'introduction de quanteurs dont la seule propriété remarquable est de s'éliminer? C'est que les quanteurs ont une grande force d'expresssion, et permettent de voir directement que certain ensembles sont constructibles; par exemple, nous dirons qu'une fonction f de K^n dans K^m est constructible si son graphe l'est; l'image d'une fonction constructible est constructible, étant définie par $(\exists\ \overline{x})\ \overline{y} = f(\overline{x})$: il en est de même de son domaine d'injectivité, défini par $(\forall\ \overline{u})\ \overline{u} = \overline{x} \vee f(\overline{u}) \neq f(\overline{x})$, etc...

Comme on le voit, la définition naturelle de nombreux ensembles constructibles met en jeu des quanteurs, dont l'élimination a un côté accidentel. Le groupe $GL_2(K)$ est constitué du sous-ensemble de K^4 défini par l'inéquation $x_{11}x_{22} \neq x_{12}x_{21}$; la multiplication matricielle a un graphe tout à fait constructible. Notons I la matrice identité, et J la matrice diagonale dont les valeurs propres sont 1 et –1; la propriété "la matrice X est conjuguée de la matrice J" se traduit naturellement par une formule avec quanteurs: $(\exists Y)(\exists Z)\ YZ = I \wedge YXZ = J$, qui devient bien une formule du langage des corps quand on remplace les matrices par des quadruplets de nombres. Par ailleurs, dans ce cas elle ne signifie rien d'autre que la matrice X admet 1 et –1 comme valeurs propres si on est en caractéristique différente de 2; en caractéristique 2, ellle signifie que $X = I$. On peut donc la remplacer par l'énoncé sans quanteurs:

$$(1 + 1 = 0 \wedge X = I) \vee (1 + 1 \neq 0 \wedge X^2 = I \wedge X \neq I \wedge X \neq -I).$$

Remarquons que le théorème de Tarski est plus précis que celui de Chevalley, puisqu'il introduit de l'uniformité: la formule $\psi(\overline{x})$ sans quanteurs équivalente à $\varphi(\overline{x})$ peut être choisie indépendemment du corps algébriquement clos considéré. Celà a une importante conséquence relative aux extensions de corps: soit $\varphi(\overline{x})$ une formule du langage des corps, soit $K \subset L$ une extension de corps, et soit $\overline{a}$ dans K; si $\overline{a}$ satisfait φ au sens de K, il n'a aucune raison en général de satisfaire φ au sens de L, puisque cette fois le champ des quanteurs de φ doit être étendu à L tout entier (par exemple, si a n'est pas carré dans K, il peut devenir carré dans L!); cependant, ce sera vrai si K et L sont algébriquement clos, puisqu'on peut alors remplacer φ par une formule ψ sans quanteurs; une telle extension, qui préserve la satisfaction des formules, est qualifiée d'<u>élémentaire</u> par les logiciens.

Par exemple, si la formule $\varphi(x,y,\overline{a})$ définit le graphe d'une fonction de K dans K, ce qui s'exprime par la satisfaction de la formule suivante par le uple de paramètres $\overline{a}$: $(\forall x)(\exists! y)\ \varphi(x,y,\overline{a})$, cette même formule définit le graphe d'une fonction de L dans L. Ou encore, si A et B sont deux matrices à coefficients dans K qui ne sont pas conjuguées dans $GL_2(K)$, elles ne le sont pas d'avantage dans $GL_2(L)$: l'extension du corps de base est sans influence sur les propriétés qui s'expriment par la satisfaction de formules du premier ordre.

Points idéaux

Le corps K étant fixé, ce qu'un logicien appelle un point est ce qu'un géomètre appele un point rationnel, c'est-à-dire un élément de K^n. Le géomètre, lui, veut prévoir toutes les extensions possibles du corps K, et ajoute systématiquement à ces points des "points idéaux", correspondant aux idéaux premiers de l'anneau de polynômes $K[\bar{X}]$; ces idéaux, le logicien les appelle types. Deux n-uples $\bar{a}$ et $\bar{b}$ d'une extension algébriquement close L de K sont dits de même type s'ils satisfont les mêmes formules à paramètres dans K; comme les quanteurs s'éliminent celà revient à dire que $\bar{a}$ et $\bar{b}$ annullent les mêmes polynômes de l'anneau $K[\bar{X}]$; comme ces polynômes forment un idéal premier de $K[\bar{X}]$, on voit bien qu'il y a correspondance entre les types et les idéaux premiers, les points rationnels s'identifiant aux idéaux maximaux de $K[X_1,...,X_n]$.

Nous dirons que le point idéal p satisfait l'équation $P(\bar{X}) = 0$ si le polynôme P est dans l'idéal associé à p; cela nous permet de considérer un ensemble constructible non seulement comme ensemble de points, mais aussi comme ensemble de points idéaux. Nous munissons ainsi l'ensemble $S_n(K)$ des idéaux premiers en n variables de la topologie constructible ayant pour base les ensembles constructibles: un ouvert est par définition une réunion d'ensembles constructibles; les constructibles y sont à la fois ouverts et fermés, et un fermé est par définition une intersection de constructibles; $S_n(K)$ est un espace totalement discontinu, et il est facile de voir qu'il est compact; ses ouverts-fermés sont exactement les ensembles constructibles.

On obtient une autre topologie $Z_n(K)$ sur ce même ensemble en prenant pour fermés les fermés de Zariski, définis par les systèmes d'équations polynômiales: c'est une topologie noethérienne; pas de suite infinie décroissante de fermés, pas de suite infinie croissante d'ouverts; naturellement, vu la dissymétrie qu'elle introduit entre l'équation et l'inéquation, la topologie de Zariski ne satisfait pas l'axiome de séparation de Hausdorff.

En ce qui concerne cette dernière, ça ne fait pas une grande différence de ne regarder que sa trace sur les points rationnels de K^n; en effet, on voit facilement qu'un constructible qui contient un point idéal contient un point rationnel, si bien que ce qu'on obtient est une topologie noethérienne dont les constructibles peuvent être identifiés avec ceux de $Z_n(K)$. Mais pour ce qui est de la compacité de $S_n(K)$, il est essentiel d'avoir ajouté les points idéaux: les points rationnels forment une partie dense, mais discrète, de $S_n(K)$.

Pour illustrer l'utilité de cette compacité, je vais déterminer les applications constructibles. Soit donc f une application constructible de K^n dans K, dont le graphe est défini par une formule à paramètres dans K. Soit p un type, et soit $\overline{a}$ un élément de ce type gisant dans une extension L de K. Il est clair qu'un automorphisme de L qui fixe (point par point) K et $\overline{a}$ doit fixer $f(\overline{a})$. Si on est en caractéristique 0, il est nécessaire que $f(\overline{a})$ s'exprime comme $R(\overline{a})$, où $R(\overline{x})$ est une fraction rationnelle à coefficients dans K; on observe que $f(\overline{x}) = R(\overline{x})$ est une formule $A(\overline{x})$ satisfaite par $\overline{a}$, c'est-à-dire par le type p. Nous recouvrons donc l'espace des types $S_n(K)$ par des ouverts-fermés; par compacité, un nombre fini d'entre eux, $A_1,\ldots,A_n$, correspondant à des fractions rationelles $R_1,\ldots,R_n$, suffit à ce recouvrement; comme, au niveau constructible, nous sommes libres de faire des combinaisons booléennes, nous pouvons supposer que $A_1,\ldots,A_n$ forment une partition de K^n: en caractéristique nulle, à toute application constructible f est associée un découpage de K^n en un nombre fini de parties constructibles sur chacune desquelles f s'exprime rationellement (avec, bien sûr, un dénominateur qui ne s'annulle pas sur l'ensemble en question). En caractéristique p, f est la racine p^{m^o} d'une telle expression.

Les logiciens travaillent au niveau constructible, qui entre dans le cadre très général de leur logique du premier ordre; ce qu'ils font est invariant par bijection constructible; il leur est difficile de trouver des propriétés intrinsèques distinguant le positif du négatif, puisqu'ils autorisent le libre emploi de la négation. Le travail des géomètres est beaucoup plus fin, et une de leur première exigence est de conserver la topologie de Zariski; ce qu'ils considèrent comme équivalents, ce sont deux variétés qui se correspondent par un isomorphisme géométrique: ces isomorphismes sont des applications constructibles très particulières.

Une des ambitions de ce recueil, c'est de montrer que certaines constructions qui sont habituellement considérées comme typiquement géométriques peuvent s'obtenir au niveau constructible; par exemple, la dimension est une notion constructible; mais surtout, lorsqu'on est en présence d'une loi de groupe, ces deux niveaux deviennent équivalents, puisque nous verrons qu'un groupe constructible est constructiblement isomorphe à un unique groupe algébrique.

Si on fait celà, ce n'est pas seulement pour des questions de méthode, mais c'est parce qu'on espère étendre ces propriétés géométriques à des contextes beaucoup plus vastes; et, en effet, dans ce qu'un logicien appelle un groupe "stable" - il s'agit d'une classe très vaste de groupes - certains sous-ensembles définissables, et certaines applications définissables ont un comportement analogue aux variétés et aux morphismes qu'on voit dans un groupe algébrique.

Celà est particulièrement vrai des groupes qui ont la propriété maximale de stabilité, dont les groupes algébriques sont un cas particulier, qui sont qualifiés de "groupes de rang de Morley fini" ; certains pensent même que ces groupes ressemblent tellement à des groupes algébriques qu'ils doivent être des groupes algébriques (à l'exception de quelques contre-exemples triviaux), et nous discuterons la conjecture suivante, due à Grégoire CHERLIN: "Tout groupe simple de rang de Morley fini est un groupe algébrique sur un corps algébriquement clos". Cette conjecture est extrêmement ambitieuse, puisqu'elle donnerait une caractérisation abstraite, sans référence à un contexte géométrique, de la notion de groupe algébrique.

Dimension

Si t est un type en n variables, c'est-à-dire un idéal premier de $K[\bar{X}] = K[X_1,...,X_n]$, le corps des quotients de l'anneau intègre $K[\bar{X}]/t$ a un certain degré de transcendance sur K, qu'on appelle le poids de t.

La hauteur (de Krull) de t est ainsi calculée par induction: elle vaut 0 si t est maximal; elle vaut au moins d+1 si t est strictement contenu dans un idéal premier de hauteur au moins d.

Le rang (de Cantor) de t est son rang de dérivation dans l'espace topologique $S_n(K)$, qui est ainsi défini: il vaut 0 si t est isolé; il vaut au moins d+1 si t est point d'accumulation de types de rang au moins d.

Les géomètres savent bien que la hauteur est égale au poids: c'est une conséquence de leur lemme de normalisation; mais la plupart d'entre eux ignorent l'existence du rang, qui est égal lui aussi au poids et à la hauteur. On donne à cette valeur commune le nom de dimension de t.

Cette dimension entre dans un cadre logique général, celui des rangs de Lascar et de Morley, qu'il est inutile de définir ici. Un simple apparté pour logiciens: pour les types en une variable, ces trois nombres sont trivialement égaux; l'inégalité de Lascar permet d'identifier RU et poids, comme d'ailleurs dans toute structure de rang 1; et dans une structure oméga-un-catégorique, il y a égalité du rang de Cantor, du rang de Morley, et du rang U de Lascar.

On appellera dimension de l'ensemble constructible (non vide) A le maximum des dimensions de ses points idéaux; grâce à la définition de la dimension par le rang, on voit que celle-ci se définit entièrement au niveau constructible: une partie A de K^n est de dimension au moins d+1 si et seulement si elle possède une infinité de sous-ensembles constructibles $A_1,\ldots,A_m,\ldots$ deux-à-deux disjoints et de dimension d; nous voyons que l'image de A par une application constructible est de dimension inférieure, que la dimension est conservée par bijection constructible.

Notons aussi que si on étend le corps K en L, on voit apparaitre de nouvelles parties constructibles de A, puisqu'il y a de nouveaux paramètres qui peuvent être coefficients de nos équations, mais elles ne font pas changer la dimension: A a même dimension au sens de K que son extension naturelle à L (si on veut, la dimension est plus une propriété de la formule qui définit A que de A lui-même, considéré comme partie de K^n).

Si A est de dimension d, il ne contient qu'un nombre fini de types de dimension d: ce nombre est appelé <u>degré de Morley</u> de A; c'est aussi le nombre maximal d'ensembles constructibles de dimension d en lequel on peut partitionner A. Plus généralement, les idéaux minimaux de A forment un ensemble discret: chacun est isolé des autres types de A par la conjonction des équations d'un de ses systèmes générateurs; par compacité de la topologie constructible, ils sont en nombre fini; mais, bien sûr, cette notion d'idéal minimal, contrairement à celle d'idéal de dimension maximale, ne se conserve pas par bijection constructible!

Elimination d'imaginaires

En géométrie comme en logique, on est très vite confronté à la prise de quotients; partant d'une structure M, on est amené à définir une autre structure N sur un ensemble de la forme M^n/E, où E est une relation d'équivalence entre n-uples d'éléments de M définissable dans M. Par exemple, la façon la plus naturelle de définir la droite projective est de l'introduire comme un quotient du plan affine privé de son origine; mais on peut également le définir en ajoutant un point à l'infini à la droite affine, si bien que, du point de vue constructible, la droite projective n'est rien que la droite affine plus un point!

C'est là un phénomène tout à fait général lorsque la structure de base est un corps algébriquement clos; les passages au quotient s'éliminent d'eux-mêmes, car, du point de vue constructible tout est affine! Par exemple, si G est un groupe constructible, et si H est un sous groupe normal définissable de G, le groupe quotient G/H est constructible; par contraste, on connait les difficultés qu'ont les géomètres débutants à faire de G/H un groupe algébrique quand G l'est.

Tout cela est conséquence du théorème suivant:

THEOREME. Toute relation d'équivalence $E(\overline{x},\overline{y})$ entre n-uples de K constructible est de la forme $f(\overline{x}) = f(\overline{y})$, où f est une application constructible de K^n dans K^m.

On voit donc qu'au lieu de parler de K^n/E il n'y a qu'à parler de $f(K^n)$, qui est une partie constructible de K^m! Pour dire la même chose de façon plus pédante: la catégorie des ensembles constructibles, avec comme morphismes les fonctions constructibles, admet des quotients.

Une classe modulo une relation d'équivalence définissable est qualifiée d'élément imaginaire, relativement à M, par les logiciens. Dans le cas des corps algébriquement clos, ces imaginaires peuvent se remplacer par des éléments réels; pour cette raison, on dit que ces corps éliminent les imaginaires.

Preuve du théorème: Je rappelle la définition, due à André Weil, du corps de définition k de l'idéal I de $K[\overline{X}]$: on considère $K[\overline{X}]/I$ comme K-espace vectoriel, et une base B de cet espace formée de monômes; tout monôme s'écrit donc de manière unique modulo I comme $\sum a_i m_i$, a_i dans K, m_i dans B et le corps k est le corps engendré par tous ces coefficients a_i. Comme on s'en rend compte aisément, un automorphisme du corps K fixe l'idéal I si et seulement s'il fixe k point par point; on en déduit que k est le plus petit corps k tel que I possède un système générateur formé de polynômes à coefficients dans k; il est donc finiment engendré, et indépendant de la base de monômes choisie.

Soit $E \subset K^n \times K^n$ une relation d'équivalence constructible; soit $\overline{a}$ un n-uple dans une extension L de K, et soit C la classe de $\overline{a}$ modulo E; soient $I_1,\ldots,I_m$ les idéaux premiers minimaux de $L[\overline{X}]$ contenus dans C, et soit $\overline{c}$ un uple engendrant le corps de définition de leur intersection I.

On voit sans peine que $I_1,\ldots,I_m$ sont les idéaux premiers minimaux contenant l'idéal radical I, si bien que, pour un K-automorphisme de L, fixer $\overline{c}$ ou permuter $I_1,\ldots,I_m$, c'est la même chose. C'est aussi la même chose que fixer C, puisque, comme E est une relation définissable avec paramètres dans

K, elle est préservée par cet automorphisme s, et $sC = C$ dès que l'intersection de C et de sC n'est pas vide.

Si en outre s fixe $\overline{a}$, il fixe C, donc $\overline{c}$; comme cela se produit pour tout $K(\overline{a})$ - automorphisme de L, ce dernier s'exprime comme $\overline{c} = f(\overline{a})$, où f est une application définissable avec paramètres dans K; cette expression de $\overline{c}$ en fonction de $\overline{a}$ ne dépend bien sûr que du type de $\overline{a}$ sur K. En utilisant la compacité de la topologie constructible, on fait un patchwork de fonctions partielles constructibles qui donne une expression uniforme du uple $\overline{c}$ de paramètres canonique de la classe C. Fin

Remarque Cette élimination des imaginaires permet de généraliser la construction de Weil, correspondant à des idéaux, c'est-à-dire des fermés de Zariski, aux ensembles constructibles les plus généraux. Pour trouver le "corps de définition" de la formule $\varphi(\overline{x},\overline{a})$, c'est-à-dire un uple $\overline{c}$ "canonique" tel que fixer $\overline{c}$ c'est la même chose que fixer la formule, on considère la relation d'équivalence constructible suivante:

$$(\forall \overline{x})\ \varphi(\overline{x},\overline{y}) \leftrightarrow \varphi(\overline{x},\overline{z}).$$

Elle est définissable sans paramètres, et se met sous la forme $f(\overline{y}) = f(\overline{z})$, où f est définissable sans paramètres; on pose $\overline{c} = f(\overline{a})$.

Variétés

J'en viens maintenant à la deuxième partie de mon programme; après avoir expliqué au géomètre ce que voit un logicien dans un corps algébriquement clos, il me faut expliquer au logicien le b.a. ba de la géométrie, c'est-à-dire expliquer ce qu'est une variété algébrique. Une variété est une certaine structure qu'on met sur un ensemble constructible; la structure de variété n'est pas conservée par bijection constructible, mais seulement par des applications constructibles très particulières, qu'on appelle les morphismes; la variété n'est donc pas déterminé par l'ensemble constructible sous-jacent; de fait, tout ensemble constructible infini peut être muni d'une infinité de structures de variété distinctes.

Ces variétés sont obtenues en recollant des morceaux d'espace affine K^n; suivant la terminologie de Dieudonné, j'appellerai ensemble affin un fermé de Zariski de K^n, c'est-à-dire une partie A de K^n défini par un système d'équations $P_1(\overline{x}) = 0 \wedge \ldots \wedge P_m(\overline{x}) = 0$. L'ensemble I des polynômes qui s'annullent sur A est l'intersection des idéaux maximaux, et aussi des idéaux premiers, qui contiennent $P_1, \ldots, P_m$: c'est l'idéal radical engendré par ces polynômes. La topologie de Zariski de A est par définition celle induite par la topologie de Zariski de K^n.

J'appelle morphisme de l'ensemble affin $A \subset K^n$ vers l'ensemble affin $B \subset K^m$ toute application f de A dans B satisfaisant au choix l'une des conditions suivantes:

(1) (définition globale) f est polynomiale, i.e. il existe un uple $(P_1(\overline{X}), \ldots, P_m(\overline{X}))$ de polynômes donnant les coordonnées de $f(\overline{X})$ lorsque $\overline{X}$ parcourt A.

(ii) (définition locale) f est localement rationelle, i.e. il existe un recouvrement $(U_i)_i$ de A par des ouverts de Zariski avec des uples $(R^i_1(\overline{X}), \ldots, R^i_m(\overline{X}))$ de fractions rationnelles, dont les dénominateurs ne s'annullent pas lorsque $\overline{X}$ parcourt U_i, et qui donne la valeur de $f(\overline{X})$ quand $\overline{X}$ est dans U_i.

On voit que ce qui fait la particularité des morphismes parmi les applications constructibles générales, c'est d'abord que les racines p°, si on est en caractéristique p, ne doivent pas intervenir, et qu'ensuite les expressions rationelles de f doivent être valables sur un ouvert de Zariski et non pas sur un constructible quelconque. Naturellement, comme les ouverts ne forment pas une algèbre de Boole, on ne peut supposer les U_i disjoints: ils auront au contraire tendance à s'intersecter très largement.

Il nous faut montrer l'équivalence des deux définitions, c'est à dire que (ii) implique (i).

Preuve : Il suffit de le faire lorsque $m = 1$; par noethérianité, les U_i sont en nombre fini, et comme chaque ouvert est réunion d'ouverts principaux, on peut supposer que chacun est défini par une seule inéquation $Q_i \neq 0$; si donc $Q_i(\overline{X}) \neq 0$, $f(\overline{X}) = P_i(\overline{X})/S_i(\overline{X})$ et $S_i(\overline{X}) \neq 0$, aucun point de A ne satisfait

$Q_i \neq 0$ et $S_i = 0$, ce qui veut dire que Q_i appartient au radical de l'idéal engendré par I et S_i, $Q_i{}^r = US_i + V$, où V est un polynôme qui s'annulle sur A; en conséquence $f = UP_i/Q_i{}^r$ lorsque $Q_i \neq 0$, soit encore $Q_i{}^r \neq 0$. On est donc ramené au cas où $f(\overline{X}) = P_i(\overline{X})/Q_i(\overline{X})$ pour $Q_i(\overline{X}) \neq 0$.

Comme aucun point de A ne peut annuller tous les $Q_i{}^2$, ces polynômes engendrent avec I l'idéal trivial, et on a une expression $1 - \Sigma B_jQ_j{}^2 = 0$ modulo I; par ailleurs $(P_iQ_j - P_jQ_i)\, Q_iQ_j$ est dans I, puisque $P_i/Q_i = P_j/Q_j$ pour chaque point de A n'annullant ni Q_i ni Q_j; en gardant i fixé, et en additionnant, on obtient modulo I:

$$\Sigma B_jP_jQ_jQ_i{}^2 = \Sigma P_iQ_iB_jQ_j{}^2 = P_iQ_i;$$

par conséquent, pour un $\overline{x}$ de A tel que $Q_i(\overline{x}) \neq 0$, $f(\overline{x}) = P_i/Q_i = \Sigma B_jP_jQ_j$; cette expression polynomiale de f est valable sur A tout entier, puisqu'elle ne dépend pas de i. Fin.

Nous appelons variété un ensemble V, recouvert par des ensembles $V_1,\ldots,V_n$, avec des bijections $f_1,\ldots,f_n$ entre $V_1,\ldots,V_n$ et des ensembles affins $U_1,\ldots,U_n$ soumis aux conditions suivantes:

- l'ensemble $U_{ij} = f_i(V_i \cap V_j)$ est un ouvert de U_i,
- l'application $f_{ij} = f_i \circ f_j{}^{-1}$, qui est une bijection de U_{ji} dans U_{ij}, est localement rationelle.

On dit souvent que la variété est un atlas dont les cartes sont les U_i; les f_{ij} sont les "applications changement de carte".

Nous définissons comme suit la topologie de Zariski de V: une partie X de V est ouverte si pour chaque i, $f_i(X \cap V_i)$ est un ouvert de U_i; comme les U_{ij} sont ouverts, et les changements de carte continus, on voit que, si X est inclus dans U_i il est ouvert si et seulement si $f_i(X)$ est un ouvert de U_i.

Etant données deux variétés $V = (V_i,f_i)$ et $W = (W_j,g_j)$, nous définissons localement la notion de morphisme de V dans W par lecture dans les cartes: $f^{-1}(W_j) \cap V_i$ est un ouvert A_{ij} de V_i, et l'application $g_j \circ f \circ f_i{}^{-1}$, restreinte à $f_i(A_{ij})$, est localement rationelle. On voit que le composé de deux morphismes est un morphisme.

Un isomorphisme, c'est un morphisme bijectif dont l'inverse est aussi un morphisme; deux variétés isomorphes ont exactement les mêmes propriétés géométriques. On dit qu'une variété est affine si elle est isomorphe à un ensemble affin, considéré comme une variété où il n'y a qu'une seule carte.

On appelle application rationelle de domaine U, où U est un ouvert de V, une application définie sur U et qui, une fois lue dans les cartes, est localement rationelle; un morphisme, c'est donc une application rationelle définie partout. Le faisceau structurel de la variété, c'est la donnée, pour chaque ouvert U de V, de l'anneau des applications rationelles de domaine U et à valeur dans K; ce faisceau détermine la variété à isomorphisme près: les géomètres préfèrent définir la variété par son faisceau, ce qui évite de recoller des cartes qui n'ont rien d'intrinsèque. C'est moins pratique pour nous, qui voulons traiter les variétés comme des objets définissables, et les manipuler comme il est usuel en logique, cest-à-dire conformément à l'intuition, en évitant tout artifice. (Le propre de la logique par rapport aux autres disciplines mathématiques, c'est de s'être donné les moyens de suivre cette intuition sans faillir à la rigueur).

Nous observons en effet qu'une variété V est bien un objet constructible: V peut être considéré comme la réunion disjointe des U_i, quotientée par la relation d'équivalence associée aux f_{ij}, qui identifie un point de U_{ij} à un point de U_{ji}; ou encore on peut faire du patchwork et poser $V = U_1 \cup (U_2 - U_{21}) \cup \ldots$ On voit alors que les morphismes deviennent des applications définissables.

Pendant qu'on y est, on remarque qu'on obtient des variétés en recollant d'autres variétés: reprendre la définition des variétés, mais en supposant cette fois que les U_i sont des variétés et non plus des ensembles affins; à chaque U_i est associé un système de cartes affines $U_{i,k}$; si on les met toutes ensemble on voit que la condition de changement de carte est bien respectée et qu'on obtient bien ainsi une variété.

Je conclus cette section par trois exemples de variété:

1 - la droite affine plus un point, sous-ensemble affin du plan défini par les équations $X(Y - 1) = 0 \wedge Y(Y - 1) = 0$,

2 - la droite avec un point dédoublé; U_1 est la droite $Y = 1$ dans le plan, U_2 la droite $Y = -1$; U_{12} est U_1 privé du point (0,1), et U_{21} est U_2 privé du point (0,-1); le changements de carte sont $f_{12}(x,-1) = (x,1)$, $f_{21}(x,1) = (x,-1)$,

3 - la droite projective; mêmes cartes que précédemment, mais avec $f_{12}(x,-1) = (1/x,1)$, $f_{21}(x,1) = (1/x,-1)$.

Du point de vue constructible, ces trois choses sont les mêmes: une droite plus un point; mais elles sont fort distinctes du point de vue géométrique; la première est affine; elle est réductible, contrairement aux deux autres; la seconde n'est pas séparable; quand à la troisième, c'est une variété complète.

<u>Variétés induites</u>

Une variété $V = (V_i, f_i)$ induit de façon naturelle une structure de variété sur certaines de ses parties constructibles. Si F est un <u>fermé</u> de V, chaque $F \cap V_i$ est un ensemble affin, et les restrictions des f_{ij} les recollent sans problème: on obtient bien une structure de variété sur F.

On en obtient également une sur un ouvert O de V; il suffit pour cela de mettre une structure de variété sur chaque $O \cap V_i$, dont la topologie de Zariski est celle induite par V, et telle que la restriction d'une application rationelle de V à $O \cap V_i$ soit encore rationelle pour cette dernière: les restriction des f_{ij} satisferont la condition de recollement.

Soit donc O un ouvert de l'ensemble affin $A \subset K^n$, défini par la disjonction $Q_1(\overline{X}) \neq 0 \vee \ldots \vee Q_s(\overline{X}) \neq 0$, et soient $O_1,\ldots,O_s$ les ouverts principaux associés à chacune de ces inéquations. Soit U_1 le fermé de $A \times K \subset K^{n+1}$ défini par l'équation $Q_1(\overline{X})Y = 1$: U_1 est bien un ensemble affin, défini par cette équation et les équations en $\overline{X}$ qui définissent A. Soit f_1 la bijection qui à un point $\overline{X}$ de O_1 associe le point $(\overline{X},1/Q_1(\overline{X}))$ de U_1: on observe que f_1, comme son inverse, sont des applications continues, l'image d'une équation en $\overline{X}$ et Y étant l'équation obtenue en substituant Y par $1/Q_1(\overline{X})$ et en chassant le dénominateur. Si donc on fait ça pour tous les O_i, les

U_{ij} sont bien des ouverts, l'application f_{ij} est bien rationelle, avec Q_i comme seul dénominateur; une application rationelle sur A se lit bien rationellement dans chaque carte U_i: nous avons tout ce qu'il nous fallait.

Chaque ouvert affine principal, grâce à cette astuce qui consiste à ajouter une coordonnée, devient ainsi une variété affine; par exemple, le groupe $GL_n(K)$, sous-ensemble de K^{n^2} défini par l'inéquation $\det(\overline{X}) \neq O$, est une variété affine.

Si nous considérons une application rationelle de domaine O_1 et à valeur dans K, par transport par f_1 cette application rationelle devient une application rationelle partout définie sur U_1, c'est-à-dire, comme nous l'avons vu, un polynôme; en revenant à O_1, on voit que cette application rationelle est polynôme en $\overline{X}$ et $Q_1(\overline{X})^{-1}$, qu'elle est de la forme $P(\overline{X})/Q_1(\overline{X})^m$, expression valable sur O_1 tout entier. Par contraste, une application rationelle dont le domaine est un ouvert non-principal peut ne pas avoir d'expression uniforme sur son domaine; et un ouvert non-principal d'une variéte affine n'est pas nécessairement affine.

Pour ces structures de variété W que nous avons définies sur les ouverts et les fermés de V, l'application identité est bien sûr un morphisme de W dans V; la topologie de Zariski de W est celle induite par celle de V; mais surtout, et c'est la raison pour laquelle on dit que W est une <u>sous-variété</u> de V, le faisceau de W est induit par celui de V, ce qui signifie (i) si r est une application rationelle de V dans K de domaine O, sa restriction à $O \cap W$ est une application rationelle de W dans K, (ii) toute application rationelle (partielle) de W dans K s'obtient localement comme restriction d'une application rationelle définie sur un ouvert de V.

C'est évident dans le cas d'un ouvert; pour un fermé, on utilise la caractérisation des fonctions rationelles ayant pour domaine un ouvert affine principal.

En combinant ces deux constructions, on obtient une structure de sous-variété sur les ensembles qui sont intersection d'un ouvert et d'un fermé; mais tous les constructibles ne sont pas ainsi (ils sont réunion d'un nombre fini d'ensembles comme ça), et on n'obtient pas de structure de variété induite sur un constructible quelconque: le problème est que le faisceau induit n'est pas nécessairement localement affine.

Si U est affine, les idéaux maximaux de l'anneau des fonctions rationelles de domaine U, c'est-à-dire des morphismes de U dans K, correspondent aux points de U; si A est le sous-ensemble de K^2 défini par $X \neq 0 \vee Y = 0$, on voit que pour le faisceau induit sur n'importe quel voisinage de (0,0) il manque toujours des points. Pourtant, si B est le fermé de K^3 défini par $XZ = 1 \vee (X= 0 \wedge Y = 0 \wedge Z = 0)$, le morphisme $(X,Y,Z) \longrightarrow (X,Y)$ de B dans K^2 définit une bijection entre B et A; on voit que l'image d'un morphisme n'est pas nécessairement une sous-variété.

<u>Remarque:</u> Tout constructible $A \subset K^n$, défini par

$$\bigvee (P_{i_1}(\overline{X}) = 0 \wedge \dots \wedge P_{i_s}(\overline{X}) = 0 \wedge Q_i (\overline{X}) \neq 0)$$

est projection du fermé $B \subset K^{n+1}$ défini par

$$\bigvee (P_{i_1}(\overline{X}) = 0 \wedge \dots \wedge P_{i_s}(\overline{X}) = 0 \wedge TQ_i (\overline{X}) = 1);$$

si on suppose en outre que les différents termes de la disjonction définissent des parties disjointes de A (et on peut toujours se ramener à ce cas, grâce à une induction aisée sur le rang et le degré de Morley!), cette projection induit une bijection de B sur A; l'inverse de cette projection n'est pas en général continu.

Par contraste, la projection d'un ouvert est toujours un ouvert; si $\varphi(\overline{x},y)$ définit un ouvert de K^{n+1}, chaque $\varphi(\overline{x},a)$ définit un ouvert de K^n; $(\exists y)\ \varphi(\overline{x},y)$, qui est la réunion des $\varphi(\overline{x},a)$, est, par noethérianité, celle d'un nombre fini d'entre eux. On peut également observer que

$$(\exists y)\ a_n(\overline{x})y^n + \dots + a_1(\overline{x})y + a_0(\overline{x}) \neq 0$$

équivaut à

$$a_n(\overline{x}) \neq 0 \vee \dots \vee a_1(\overline{x}) \neq 0 \vee a_0(\overline{x}) \neq 0.$$

Points génériques

Une variété V est dite irréductible si elle ne peut s'écrire comme réunion de deux sous-fermés propres; une application aisée du lemme de König montre que V s'écrit sous la forme d'une réunion finie $V_1 \cup \ldots \cup V_n$ de fermés irréductibles; si on jette les V_i qui sont contenus dans un autre, on observe que ceux qui restent sont les fermés irréductibles maximaux de V: on les appelle composantes irréductibles de V; la décomposition de V alors obtenue est unique. Les composantes connexes de V sont également en nombre fini: on les obtient en regroupant les composantes irréductibles qui s'intersectent, et en continuant de proche en proche. Si A est un sous-ensemble affin de K^n, il est irréductible si et seulement si l'idéal I associé est premier; dans le cas général, les composantes irréductibles de A correspondent aux idéaux premiers minimaux contenant l'idéal radical I.

Si V est irréductible, l'intersection de deux ouverts non-vides est non-vide; tout ouvert non-vide est dense dans V; il est également irréductible pour la topologie de Zariski induite. On voit qu'il existe un unique type complet satisfaisant toutes les formules définissant un ouvert de V; en lisant la variété dans une carte affine, on voit que c'est l'unique type de dimension maximale dans V; on l'appelle point générique de V. Si V est réductible, on appelle parfois points génériques de V ceux de ses composantes irréductibles: nous éviterons cette terminologie.

Si V est irréductible, tout ensemble constructible contenant le générique, c'est-à-dire toute partie constructible de V de même dimension que V, contient un ouvert non-vide. En effet cet ensemble s'écrit

$$(F_1 \cap O_1) \cup \ldots \cup (F_n \cap O_n);$$

le générique appartient par exemple à $F_1 \cap O_1$; comme il appartient au fermé F_1, par définition de la généricité $F_1 = V$, et notre ensemble contient l'ouvert O_1.

Supposons que le corps de base soit de caractéristique nulle, et considérons une fonction constructible f de V dans W, V étant une variété irréductible; nous restreignons f à une carte affine V_1, et nous savons qu'il est possible de découper V_1 en un nombre fini de parties constructibles sur chacune desquelles f a une expression rationelle; celle de ces parties qui contiennent le générique contient un ouvert O: nous voyons que la restriction de f à O est une application rationelle, f est "génériquement" un morphisme! Si f est une bijection constructible entre les variétés irréductibles V et W, elle induit ce que les géomètres appellent une "équivalence birationelle" entre V et W, c'est-à-dire un isomorphisme entre un ouvert de V et un ouvert de W.

En caractéristique p, les racine p° s'introduisent dans le paysage.

Soit V une variété irréductible; considérons deux applications rationelles f_1 et f_2 de V dans K de domaines respectifs O_1 et O_2; on voit en lisant la chose dans des sous-ouverts affines de $O_1 \cap O_2$, que la condition $f_1(\overline{x}) \neq f_2(\overline{x})$ définit un ouvert de cette intersection (c'est vrai en général si la variété d'arrivée est <u>séparée</u>; voir ci-après pour la définition); si donc f_1 et f_2 agréent génériquement, c'est-à-dire sont égales sur un ouvert non-vide (on dit qu'elles ont même germe), elles sont égales sur toute l'intersection $O_1 \cap O_2$; on voit qu'à chaque germe correspond une fonction rationelle dont le domaine de définition est maximal: elle ne peut se prolonger en une fonction rationelle sur un ouvert plus large. On observe également que deux morphismes de V dans K (ou plus généralement dans une variété séparée) et qui sont égaux sur un ouvert sont égaux partout.

Une dernière utilisation des ensembles irréductibles: soit maintenant V une variété quelconque; nous observons d'abord que si F est un fermé irréductible de V, et O un ouvert de V, alors $F \cap O$ est un fermé irréductible de O; par conséquent, si F_1 est un fermé irréductible strictement contenu dans le fermé irréductible F_2, la dimension de F_1 est strictement inférieure à celle de F_2: on se ramène au cas affine, où on utilise la caractérisation de la dimension par la hauteur de Krull.

Soit alors X un constructible de V; pour tout type t de X, le plus petit fermé F_t de V contenant t est irréductible, et t est son générique; comme les dimensions sont bornées, chaque F_t est contenu dans un $F_{t'}$ maximal; on observe que, pour la topologie constructible, les t' en question forme une partie discrète de X; par compacité, ils sont en nombre fini, soient

$t_1,\ldots,t_n$, et $F_{t_1} \cup \ldots \cup F_{t_n}$ est l'adhérence $\overline{X}$ de X pour la topologie de Zariski; on voit qu'elle est de même dimension que X, et que tous les points de dimension maximale de $\overline{X}$ sont dans X: en conclusion X et $\overline{X}$ ont même rang et même degré de Morley, la dimension de $\overline{X} \setminus X$ est strictement inférieure à celle de X.

Produits de variétés

Si $A \subset K^n$ est défini par les équations $P_i(\overline{x}) = 0$, et $B \subset K^m$ par les équations $Q_j(\overline{y}) = 0$, le produit $A \times B \subset K^{n+m}$ est défini par les équations $\ldots \wedge P_i(\overline{x}) = 0 \wedge \ldots \wedge Q_j(\overline{y}) = 0 \wedge \ldots$; le produit de deux ensembles affins est bien leur produit cartésien si on les considère comme ensembles de points de K^n, K^m, K^{n+m}; à l'exception du cas où l'un d'entre eux est fini, il n'en est plus de même si on les considère comme ensembles de points idéaux puisque la donnée du type de $\overline{x}$ et du type de $\overline{y}$ ne suffit pas à déterminer le type de $\overline{x}\frown\overline{y}$; et la topologie de Zariski de $A \times B$ est strictement plus fine que le produit des topologies de Zariski de A et de B, une équation en $\overline{x}\frown\overline{y}$ ne pouvant en général se ramène à un système d'équations en $\overline{x}$ seulement et d'équations en $\overline{y}$ seulement.

Si $V = (V_i,f_i)$ et $W = (W_j,g_j)$, on définit sur le produit $V \times W$ une structure de variété en prenant pour cartes les produits $V_i \times W_j$; on vérifie que ça marche, que ça a bien les propriété fonctorielles d'un produit, et que ça se comporte bien par rapport aux variétés induites.

On dit que la variété V est séparée si la diagonale - ensemble des points $(\overline{x},\overline{x})$ de $V \times V$ - est un fermé de $V \times V$; dans le cas d'espaces topologiques, cette condition équivaudrait à la séparation de Haussdorf. Une variété affine est séparée, puisque l'égalité $\overline{x} = \overline{y}$ s'exprime par des équations. Certains réservent le terme "variété" aux variété séparées, et appellent prévariétés les autres.

Une variété V est dite complète si pour toute variété W la deuxième projection $V \times W \to W$ est fermée (i.e. l'image d'un fermé est fermée). Pour jeter un peu de lumière sur cette définition, vous pouvez observer que, dans le cas des espaces topologiques, cette propriété de diagramme serait celle des compacts (anglais). Une sous-variété fermée d'une variété complète est complète; l'image d'une variété complète par un morphisme arrivant dans une variété séparée est un fermé: c'est donc une variété complète; un produit de variétés complètes est une variété complète. Il y a incompatibilité d'humeur entre les variétés complètes et les variétés affines: une variété complète et affine est finie.

L'exemple le plus célèbre de variétés complètes, ce sont les espaces projectifs, et leurs fermés; elles sont essentielles pour établir les propriétés des borels d'un groupe linéaire, que je n'aborderai pas en détail: je renvoie donc aux ouvrages spécialisés.

Groupes algébriques

Un groupe algébrique est un groupe constructible G, qui est une variété, telle que la loi de groupe soit un morphisme de $G \times G$ dans G. On trouvera dans la littérature la justification des affirmations suivantes:

1- La variété G est séparée (c'est facile: la diagonale D a même dimension que G; si $(a,a') \in \bar{D} \setminus D$, il en de même de chaque (ba,ba'), et $\bar{D} \setminus D$ aurait même dimension).

2 - Tout sous-groupe définissable de G est fermé (même preuve).

3 - La composante connexe $G°$ de l'identité est un sous-groupe fermé de G; c'est le plus petit sous-groupe fermé de G, et il est normal dans G; il est également irréductible, et les composantes irréductibles de G sont les classes modulo $G°$; il y a exactement un type de dimension maximale par classe modulo $G°$.

4 - Un constructible A de G est de dimension maximale si et seulement s'il est d'intérieur non-vide si et seulement si G est recouvert par un nombre fini de translatés, à droite ou à gauche, de A : $G = a_1A \cup \ldots \cup a_nA =$

$Ab_1 \cup \ldots \cup Ab_m$; une telle partie est dite <u>générique</u>. Ces parties de dimension maximale se caractérisent donc simplement grâce à la loi de groupe, sans qu'il soit besoin de faire intervenir la dimension! Nous verrons dans ce recueil que ces parties génériques ont un analogue dans le cas général des groupes stables: on y dispose de l'analogue des constructibles Zariski-denses (ceux dont le complémentaire n'est pas générique), bien qu'on y ait rien qui corresponde à la topologie de Zariski.

Si X est constructible et Zariski-dense (si G est connexe, cela signifie que X est d'intérieur non vide) tout élément de G s'écrit comme produit de deux éléments de X.

5 - Si H est un sous-groupe fermé de G, on peut définir sur l'ensemble G/H des classes à droite modulo H une structure de variété telle que l'action de G par multiplication à gauche soit un morphisme de $G \times G/H$ dans G/H. Quand H est normal dans G, G/H devient ainsi un groupe algébrique.

6 - Si G/H est une variété complète, on dit que H est un sous-groupe <u>parabolique</u> de G; il existe un plus petit sous-groupe parabolique G_a normal dans G, qui est contenu dans G°; c'est aussi le plus grand sous-groupe fermé affine connexe contenu dans G. Les sous-groupes paraboliques minimaux de G sont aussi les sous-groupes fermés connexes résolubles maximaux de G_a; ils sont tous conjugués dans G_a; on les appelle groupes de Borel de G^a.

7 - Un groupe algébrique connexe et complet est appelé <u>variété abélienne</u>; c'est un groupe commutatif; s'il n'est pas réduit à l'identité, il contient des éléments de tout ordre premier à la caractéristique.

Tout groupe algébrique G contient une plus grande sous-variété abélienne A, qui est centrale dans G, mais le quotient G/A n'est pas affine en général.

8 - Si Z(G) est le centre de G, G/Z(G) est affine; un groupe algébrique sans centre, en particulier un groupe algébrique simple est affine.

9 - Tout groupe algébrique affine G est linéaire, c'est-à-dire qu'il est isomorphe - par un isomorphisme qui est à la fois isomorphisme de groupe et variété - à un sous-groupe fermé d'un $GL_n(K)$ (ouvert principal de K^n défini par det $(\bar{X}) \neq 0$, muni de la multiplication matricielle, qui est polynomiale); G a beaucoup de telles représentations linéaires, mais un certain nombre de choses, en particulier la décomposition d'un élément comme produit d'un élément semi-simple et d'un élément unipotent, sont indépendantes de la représentation.

10 - On connait beaucoup de choses de la structure d'un groupe affine quand on connait sa décomposition en classes doubles modulo un borel; un élément central dans un borel de G est central dans G: les borels d'un groupe simple n'ont pas de centre.

Références

Logique générale:

Bruno POIZAT, Cours de Théorie des Modèles
Nur alMantiq walMa'arifah, Villeurbanne, 1985.

Logique et corps algébriquement clos:

Chantal BERLINE, Déviation des types dans les corps algébriquement clos, in Théories Stables III, I.H.P., Paris, 1983.

Elimination des imaginaires:

Bruno POIZAT, Une théorie de Galois imaginaire
J.S.L. 48, 1983, 1151-1170.

Géométrie algébrique générale:

Jean DIEUDONNE, Cours de Géométrie Algébrique
P.U.F., Paris, 1974.

Groupes algébriques:

Ellis KOLCHIN, Differential Algebra and Algebraic Groups
Academic Press, New York, 1973.

Groupes algébriques affines:

T.A. SPRINGER, Linear Algebraic Groups
Birkhauser, Basel, 1981

James E. HUMPHREYS, Linear Algebraic Groups
Springer, New York, 1975

Tout en un seul ouvrage:

Bruno POIZAT , Groupes Stables
Nur alMantiq walMa'arifah, Villeurbanne, 1987

Bruno POIZAT
82, rue Racine
69100 Villeurbanne
France

COUNTABLY CATEGORICAL EXPANSIONS OF PROJECTIVE SPACES

Simon Thomas[1]

1. INTRODUCTION

A number of important problems in model theory ask what extra structure can be imposed upon a model M, while preserving various model-theoretic properties of M. For example, it has been conjectured that if extra structure is imposed upon an algebraically closed field F, then the resulting model F^+ no longer has finite Morley rank. In this paper, we shall discuss various open problems concerning ω-categorical structures of the form $M = \langle PG(\omega,q),R\rangle$. Here $PG(\omega,q)$ denotes an infinite dimensional projective space over the finite field $GF(q)$ and R is some extra relation. Our starting point is the observation that structures of this form provide an interesting test case for Lachlan's conjecture that a stable ω-categorical structure is ω-stable.

Theorem 1.1

Suppose that $M = \langle PG(\omega,q),R\rangle$ is ω-stable and ω-categorical. If $G = \text{Aut } M$ acts primitively on M, then M is strictly minimal.

Proof

By [8], M can be expressed as a union of finite algebraically closed subsets, $M = \bigcup_{i\in\omega} M_i$, such that

(i) $G_i = \text{Aut } M_i$ acts primitively on M_i ;

(ii) G_i has the same number n_2 of orbits on the lines of M_i as G has on the lines of M. Let $M_i = \langle P_i, R_i\rangle$, where P_i is a subspace of dimension d_i. (Throughout this paper, we will be using vector space dimension; so that

[1]Research partially supported by NSF grant DMS-8703229

points are 1-dimensional, lines are 2-dimensional, etc.) We can suppose that $d_0 > 6$. By Hering [10], for each i either $n_2 = 1$ and $PSL(d_i,q) \leq G_i$, or else $n_2 = 2$, $PSp(d_i,q) \leq G_i$ and G_i preserves a nondegenerate symplectic polarity of P_i. (A statement of Hering's theorem can be found in Section 4). If the former occurs for all $i \in \omega$, then M is clearly strictly minimal. On the other hand, if $PSp(d_i,q) \leq G_i$ for all $i \in \omega$, then it is easily seen that G preserves a nondegenerate symplectic polarity of $PG(\omega,q)$, which contradicts the assumption that M is stable. □

Exercise 1.2

Find an elementary proof of this result, using the coordinatization theorem. (The proof of Hering's theorem makes use of the classification of the finite simple groups).

This paper is organized as follows. In Section 2, we shall discuss various conditions which imply that a 2-transitive stable ω-categorical structure has the form $\langle PG(\omega,q),R\rangle$. Sections 3 and 4 consider algebraic closure in such structures. Finally in section 5, we discuss projective space versions of results of Cameron [5], [6]. In particular, we will give a characterization of infinite dimensional symplectic spaces over finite fields.

If (G,Ω) is a permutation group and $X \subseteq \Omega$, then $G_{\{X\}}, G_{(X)}$ denote the setwise and pointwise stabilizers of X in G. The stabilizer of a point $\alpha \in \Omega$ is written as G_α.

A linear space is a structure $S = (\Omega,\mathcal{L})$, where $\mathcal{L} \subseteq P(\Omega)$, such that

(i) every pair of points $\alpha, \beta \in \Omega$ lie in a unique element of $\mathcal{L}$,

(ii) if $\ell, \ell' \in \mathcal{L}$ then $|\ell| = |\ell'| > 2$.

The elements of $\mathcal{L}$ are called lines.

If P is a projective space, then $P^{(k)}$ denotes the set of k-dimensional subspaces of P. $PG_2(\omega,q)$ is the linear space $(PG(\omega,q), PG^{(2)}(\omega,q))$. We use a similar notation for affine spaces.

2. TRIANGLE TRANSITIVE LINEAR SPACES

The results in this section give conditions under which a 2-transitive stable ω-categorical structure has the form $<PG(\omega,q),R>$ or $<AG(\omega,q),R>$.

Theorem 2.1

Let M be a stable ω-categorical structure and let $G = \text{Aut } M$. Suppose that

(i) G acts 2-transitively on M;

(ii) if $\alpha \neq \beta \in M$, then $|\text{acl}(\alpha,\beta)| > 2$;

(iii) $G_{\alpha,\beta}$ acts transitively on $M \setminus \text{acl}(\alpha,\beta)$.

Let $\mathcal{L} = \{\text{acl}(\alpha,\beta) \mid \alpha \neq \beta \in M\}$. Then $(M,\mathcal{L})$ isomorphic to $PG_2(\omega,q)$ for $q \geq 2$ or $AG_2(\omega,q)$ for $q \geq 3$.

Proof

Since M is 2-transitive and ω-categorical, any two points $\alpha \neq \beta \in M$ lie in a unique element of $\mathcal{L}$ and each element of $\mathcal{L}$ has the same finite cardinality. Thus $(M,\mathcal{L})$ is a linear space. By (iii), G acts transitively on the triangles of M. Let $\{\alpha_1,\alpha_2,\alpha_3\}$ be a triangle, and let P be the plane generated by $\{\alpha_1,\alpha_2,\alpha_3\}$. If $\{\beta_1,\beta_2,\beta_3\}$ is a second triangle of P, then there exists $\pi \in G$ such that $\beta_i^{\pi} = \alpha_i$ for $1 \leq i \leq 3$. Since $\{\alpha_1,\alpha_2,\alpha_3\} \subseteq P \cap P^{\pi}$, we have that $P^{\pi} = P$. Hence $H = \text{Aut } P$ acts transitively on the triangles of P. By Kantor [11], P must be one of the following linear spaces:

(a) $PG(2,q)$ for some $q \geq 2$;

(b) $AG(2,q)$ for some $q \geq 3$;

(c) the unital U associated with $PSU(3,4)$.

To eliminate (c), we make use of the stability of M. Since M is 2-transitive, the unique type $p \in S_1(\varnothing)$ is stationary. By (iii), $tp(\sigma|\alpha,\beta)$ doesn't fork over $\varnothing$ for all $\sigma \in M \setminus \text{acl}(\alpha,\beta)$. Hence $tp(\sigma|\text{acl}\,(a,\beta))$ doesn't fork over $\varnothing$. It follows that if $\ell \in \mathcal{L}$, then $G_{(\ell)}$ acts transitively on $M \setminus \ell$. Arguing as in the first paragraph, for each line ℓ of P, $H_{(\ell)}$ acts transitively on $P \setminus \ell$. Now consider the unital U associated with $PSU(3,4)$. By O'Nan

[14], $H = \text{Aut } U = P\Gamma U(3,4)$. It is easily checked that H acts sharply transitively on the triangles of U. Since $H_{\alpha,\beta}$ acts nontrivially on the line ℓ containing $\alpha \neq \beta \in U$, $H_{(\ell)}$ isn't transitive on $U \setminus \ell$.

It is well known that if each plane P of $(M,\mathcal{L})$ is isomorphic to PG(2,q), then $(M,\mathcal{L}) \simeq PG_2(\omega,q)$. By Buekenhout [2], if each plane is isomorphic to AG(2,q) for some $q \geq 4$, then $(M,\mathcal{L}) \simeq AG_2(\omega,q)$. The analogue of Buekenhout's theorem is false for $q = 3$. (For example, see Young [20]). However, by [19], if $(M,\mathcal{L})$ is a triangle transitive linear space in which each plane is isomorphic to AG(2,3), then $(M,\mathcal{L}) \simeq AG_2(\omega,3)$. □

There are two ways in which Theorem 2.1 should be strengthened.

Problem 2.2

Show that the conclusion still holds without the assumption that M is stable. In other words, prove that there is no ω-categorical triangle transitive linear space in which each plane is isomorphic to the unital U associated with PSU(3,4).

Problem 2.3

Show that the conclusion still holds without hypothesis (iii).

The next result shows that to solve problem 2.3, it is enough to show that each plane is projective or affine.

Theorem 2.4

Let $S = (\Omega,\mathcal{L})$ be a 2-transitive stable ω-categorical linear space. If each plane of S is isomorphic to a finite projective or affine plane, then either $S \simeq PG_2(\omega,q)$ for $q \geq 2$ or $S \simeq AG_2(\omega,q)$ for $q \geq 3$.

Proof

By Teirlinck [16], either all planes of S are projective or all planes of S are affine. The only difficulty occurs when each plane P of S is an affine plane in which the lines have cardinality 3. (The results which we used in the proof of Theorem 2.1 do not assume that the planes are Desarguesian). In this case, $P \simeq AG(2,3)$. It is possible to coordinatise S by a commutative Moufang loop of exponent 3 [20]. First we make S into a quasigroup by defining a binary operation o as follows:

$x \circ x = x$

$x \circ y = z$ for $x \neq y$ if $\{x,y,z\}$ is a line.

Now fix a point $e \in S$. Then the loop of S based at e, Q, is defined by

$xy = (e \circ x) \circ (e \circ y).$

Q is a commutative loop of exponent 3 with identity element e. Furthermore, Q satisfies the Moufang condition

$(xy)(zx) = (x(yz))x.$

Now $S \simeq AG(\omega,3)$ if and only if Q is a group, i.e., the associative law also holds. Suppose then that Q is not a group. Notice that any automorphism of S which fixes e induces an automorphism of Q. Hence Aut Q acts transitively on $Q \setminus e$.

A subloop H of Q is said to be normal if for all $x,y \in Q$

$$xH = Hx, \quad (Hx)y = H(xy), \quad y(xH) = (yx)H.$$

By Bruck 11.1 [1], Q is not simple. Since Aut Q acts transitively on $Q \setminus e$, this implies that Q has no minimal nontrivial normal subloops. (See Bruck 8.1 [1]). Let $a \in Q \setminus e$ and let $N(a)$ be the smallest normal subloop containing a. Since Q is ω-categorical, $N(a)$ is a -definable. Let $e \neq M \lneq N(a)$ be a normal subloop. Then if $b \in M \setminus e$, we have that $N(b) \lneq N(a)$. Since a,b lie in the same Aut Q-orbit, this means that Q is unstable. Hence S is unstable. □

3. ALGEBRAIC CLOSURE

Throughout this section, we will suppose that $M = \langle PG(\omega,q), R\rangle$ satisfies the following conditions.

3.1 $G = \mathrm{Aut}\ M$ acts 2-transitively on M.

3.2 M is stable ω-categorical, but not ω-stable.

We begin by collecting together some elementary lemmas. If $A \subseteq M$, then $\langle A \rangle$ denotes the subspace of $PG(\omega,q)$ generated by A.

Lemma 3.1

The unique type $p \in S_1\emptyset$ is stationary. □

Lemma 3.2

Let $\{a_i \mid 1 \leq i \leq n\}$ be a Morley sequence for $p \in S_1(\emptyset)$ and let $X = \langle a_i \mid 1 \leq i \leq n \rangle$. Then $PSL(X) \leq G_{\{X\}} / G_{(X)}$.

Proof

Let $a_n \neq \beta \in \langle a_1,\ldots,a_n \rangle \setminus \langle a_1,\ldots,a_{n-1} \rangle$. Then the line $\langle \beta, a_n \rangle$ intersects the hyperplane $\langle a_1,\ldots,a_{n-1} \rangle$ in a point α. Suppose that $tp(\beta \mid \langle a_1,\ldots,a_{n-1} \rangle)$ forks over α. Then, by Shelah III 6.7 [15], $tp(a_n \mid \langle a_1,\ldots,a_{n-1} \rangle)$ forks over α, a contradiction. As G is 2-transitive and $tp(\beta \mid \langle a_1,\ldots,a_{n-1} \rangle)$ doesn't fork over α, it follows that $tp(\beta \mid \langle a_1,\ldots,a_{n-1} \rangle)$ doesn't fork over $\emptyset$.

Let $H = G_{\{X\}}/G_{(X)}$. For each $1 \leq i \leq n$, let $Y_i = \langle \{a_1,\ldots,a_n\} \setminus \{a_i\} \rangle$. By the previous paragraph, $H_{(Y_i)}$ acts transitively on $X \setminus Y_i$, i.e., $X \setminus Y_i$ is a Jordan subset for H. It follows that

$$\bigcup_{i \neq 1} (X \setminus Y_i) = X \setminus \bigcap_{i \neq 1} Y_i = X \setminus \{a_1\}$$

is a Jordan subset. (For example, see Neumann [13]). Similarly $X \setminus \{a_2\}$ is a Jordan subset, and so H acts 2-transitively on X. Clearly we can suppose that $n \geq 5$. The result now follows by Cameron and Kantor [7]. □

Hence, by passing to a suitable quotient geometry, we can suppose that M also satisfies the following condition.

3.3 G has more than one orbit on the set $PG^{(3)}(\omega,q)$ of planes of M.

For $\alpha \neq \beta \in M$, define

$$\varphi(M; \alpha,\beta) = \{\gamma \in M \mid tp(\gamma \mid \alpha,\beta) \text{ forks over } \emptyset\}.$$

Then $\langle \alpha,\beta \rangle \subsetneq \varphi(M; \alpha,\beta)$.

$RM(-)$ denotes Morley rank.

Lemma 3.3

$RM(\varphi(M; \alpha,\beta)) = \infty$.

Proof

Suppose not. First suppose that $\varphi(M; \alpha,\beta)$ is finite. Then $\varphi(M; \alpha,\beta) = acl(\alpha,\beta) \supsetneq \langle\alpha,\beta\rangle$, and $G_{\alpha,\beta}$ acts transitively on $M \setminus acl(\alpha,\beta)$. Let $\mathcal{L} = \{acl(\alpha,\beta) \mid \alpha \neq \beta \in M\}$. By Theorem 2.1, $(M,\mathcal{L})$ is isomorphic to $PG_2(\omega,r)$ or $AG_2(\omega,r)$ for some prime power $r \neq q$. If X is a finite algebraically closed subset of M, then X must be a subspace of both $PG(\omega,q)$ and $(M,\mathcal{L})$. Suppose that $(M,\mathcal{L}) \simeq PG_2(\omega,r)$. If $\ell \in \mathcal{L}$, then

$$|\ell| = r + 1 = q^n + q^{n-1} + ... + q + 1$$

for some $n \geq 2$. But this means that $r = q^n + ... + q$ is not a prime power, a contradiction. Hence $(M,\mathcal{L}) \simeq AG_2(\omega,r)$. So if $\ell \in \mathcal{L}$, then $|\ell| = r = q^{n+1} - 1/q-1$ for some $n \geq 2$. Let X be a finite algebraically closed subset of M, chosen so that

$$|X| = r^d = q^{m+1} - 1/q-1$$

for some $m + 1 > \max\{n+1,6\}$. By Zsigmondy [21], there exists a prime p such that $p | q^{m+1} - 1$ but p does not divide $q^i - 1$ for any $1 \leq i \leq m$. Again this contradicts the fact that r is a prime power.

Thus $0 < RM(\varphi(M; \alpha,\beta)) < \infty$. Let $L(\alpha,\beta) = \varphi(M;\alpha,\beta)$. By Shelah V 7.8 [15], $\bigcup_{\gamma\neq\delta\in L(\alpha,\beta)} L(\gamma,\delta)$ also has Morley rank less than ∞, and so $L(\alpha,\beta) = \bigcup_{\gamma\neq\delta\in L(\alpha,\beta)} L(\gamma,\delta)$. In particular, if $\gamma \neq \delta \in L(\alpha,\beta)$, then $L(\gamma,\delta) \subseteq L(\alpha,\beta)$. Since M is 2-transitive, we must have that $L(\gamma,\delta) = L(\alpha,\beta)$. Let $\mathcal{L} = \{L(\alpha,\beta) \mid \alpha \neq \beta \in M\}$. Then $(M,\mathcal{L})$ is a linear space, in which each line is infinite. Let α, β, γ be a noncollinear triple of $(M,\mathcal{L})$. Define a sequence of $\{\alpha,\beta,\gamma\}$-definable subsets of M inductively by

$$S_0 = L(\alpha,\beta) \cup L(\beta,\gamma) \cup L(\gamma,\alpha)$$
$$S_{n+1} = \bigcup_{a\neq b\in S_n} L(a,b).$$

There exists n such that $S_n = S_{n+1}$. Clearly $RM(S_n) < \infty$. But now it is easily checked that $(S_n, \{L(a,b) \mid a \neq b \in S_n\})$ is a pseudoplane, contradicting [8]. □

We can now easily prove

Theorem 3.4

There exists a finite dimensional subspace $X \subseteq M$ such that $X \subsetneq \mathrm{acl}(X)$.

My reason for recording this result is that it is conceivable that this already leads to a contradiction for the following problem still seems to be open.

Problem 3.5

Does there exist a (not necessarily stable) 2-transitive ω-categorical expansion of $PG(\omega,q)$ in which $X \subsetneq \mathrm{acl}(X)$ for some finite dimensional subspace X?

Proof of Theorem 3.4

Let Δ be the Booleanly closed set of formulas generated by $\{\varphi(x;\alpha,\beta) \mid \alpha \neq \beta \in M\}$. Choose $\pi(x;\bar{c}) \in \Delta$ such that

(i) $RM(\pi(x;\bar{c})) = \infty$;

(ii) $R_\varphi(\pi(x;\bar{c})) = k$ is minimal subject to (i) and $\deg_\varphi(\pi(x;\bar{c})) = 1$.

By Lachlan [12], we can also suppose that

(iii) $\pi(x;\bar{z})$ is normal with respect to $\varphi(x;\bar{y})$.

By Lemma 3.3, $\pi(M;\bar{c}) \neq M$. Let $\mathfrak{L}$ be the set of conjugates of $\pi(M;\bar{c})$. Then if $A \neq B \in \mathfrak{L}$, $RM(A \cap B) < \infty$.

Suppose that there exist $A,B \in \mathfrak{L}$ such that $0 < RM(A \cap B) < \infty$. Then we can replace $\mathfrak{L}$ by the set of conjugates of a normal strongly minimal subset of M.

Hence we can assume that if $A \neq B \in \mathfrak{L}$, then $A \cap B$ is finite. Let $n \in \mathbb{N}$ be such that if $A \neq B \in \mathfrak{L}$, then $|A \cap B| < n$. Since G acts primitively on M, there exist $A \neq B \in \mathfrak{L}$ with $A \cap B \neq \emptyset$. Let $\gamma \in A \cap B$. Define inductively $\alpha_i \in A$, $1 \leq i \leq n$, so that $\alpha_{i+1} \notin <\alpha_1, \ldots, \alpha_i, \gamma>$. Then $\gamma \notin <\alpha_1, \ldots, \alpha_n>$. Continuing in this fashion, we can find $\beta_i \in B$, $1 \leq i \leq n$, such that $\gamma \notin X = <\alpha_i, \beta_i \mid 1 \leq i \leq n>$. But clearly $\gamma \in acl(X)$.

□

The situation described in the second paragraph of the above proof seems extremely unlikely. More generally, the following problem may be manageable.

Problem 3.6

Prove that if M is a primitive stable ω-categorical structure with an infinite definable ω-stable subset, then M is ω-stable.

We can suppose that for each $A \in \mathfrak{L}$, $G_{\{A\}}$ acts transitively on A. Let $S \subset A$ be a subset of maximal cardinality, subject to the condition that S lies in infinitely many elements of $\mathfrak{L}$. (It is easily seen that each $\alpha \in M$ lies in infinitely many elements of $\mathfrak{L}$). Let $\Omega = \{S^\pi \mid \pi \in G\}$ and define an incidence relation I by

$$T \ I \ B \iff T \subset B$$

for $T \in \Omega$ and $B \in \mathfrak{L}$. Then $(\Omega, \mathfrak{L})$ is a pseudoplane. Thus Theorem 3.4 is just the observation that it is possible to define a pseudoplane geometry in M^{eq} which interacts nontrivially with the projective geometry on M.

Given the nature of our hypotheses on M, it seems best to continue working directly with $(M, \mathfrak{L})$. At this point, we would like to have that if $A \in \mathfrak{L}$, then $<A>$ is a proper subspace of M. (Of course, this is true if A is strongly minimal). Unfortunately this does not follow immediately from general facts about forking.

Problem 3.7

Let $\mathfrak{L}$ be a uniformly definable family of infinite almost disjoint subsets of the (not necessarily stable) 2-transitive ω-categorical structure $M = <PG(\omega,q),R>$. Is $<A>$ a proper subspace of M for $A \in \mathfrak{L}$?

Suppose that $\langle A \rangle$ is indeed a proper subspace of M. By the stable descending chain condition for uniformly definable subspaces of M, there exists a definable subspace $S \leq \langle A \rangle$ such that the set S of conjugates of S is almost disjoint. Showing that such a family S cannot exist is a very interesting geometric problem. The special case when $(PG(\omega,q),\mathsf{S})$ is actually a linear space seems particularly attractive, although still very difficult.

4. DESIGNS OVER FINITE FIELDS

Again $M = \langle PG(\omega,q),R \rangle$ satisfies conditions 3.1 to 3.3. It seems useful to split the analysis of M into 3 cases, depending on the action of G_α on $\mathbb{P}_\alpha = PG(\omega,q)/\langle\alpha\rangle$. The possibilities are:

(A) G_α preserves a nontrivial equivalence relation on $\mathbb{P}_\alpha$ which has finite classes;

(B) G_α acts primitively on $\mathbb{P}_\alpha$;

(C) G_α preserves a nontrivial equivalence relation on $\mathbb{P}_\alpha$ which has infinite classes.

The cases are listed in what I believe to be the order of difficulty. In this section, case A will be discussed. This corresponds to the situation when $\ell \subsetneq \mathrm{acl}(\ell)$ for $\ell \in PG^{(2)}(\omega,q)$. Let $\mathcal{D} = \{\mathrm{acl}(\ell) \mid \ell \in PG^{(2)}(\omega,q)\}$. Then $(M,\mathcal{D})$ is a linear space. Let $\dim Q = m > 2$ for $Q \in \mathcal{D}$.

Definition 4.1

Given a finite field $\mathbf{F} = GF(q)$, a t–(n,m,λ) design over $\mathbf{F}$ is an incidence structure $\mathcal{D} = (P,\mathcal{D})$ which satisfies the following conditions:

(a) P is an n-dimensional projective space over $\mathbf{F}$.

(b) $\mathcal{D} \subseteq P^{(m)}$.

(c) Each $S \in P^{(t)}$ lies in exactly λ elements of $\mathcal{D}$.

If $\mathcal{D} = P^{(m)}$, then the design $\mathcal{D}$ is said to be trivial. We define $\mathrm{Aut}\,\mathcal{D} = \{\pi \in P\Gamma L(n,q) \mid \mathcal{D}^\pi = \mathcal{D}\}$.

Let X be a finite algebraically closed subspace of M with $\dim X = n > m$, and let $\mathcal{D}_X = \{Q \in \mathcal{D} \mid Q \subseteq X\}$. Then $(X,\mathcal{D}_X)$ is an example of a nontrivial 2-(n,m,1) design over GF(q). Despite the efforts of a number of finite combinatorialists, no such designs have been found. In fact, the only known examples of nontrivial 2-designs over finite fields are those in the infinite family described in [17]. For each of these designs, $\lambda = 7$.

Lemma 4.2

Let $(P,\mathcal{D})$ be a 2-(n,m,1) design over GF(q).

(i) Each point $\alpha \in P$ lies in $r = q^{n-1} - 1/q^{m-1} - 1$ elements of $\mathcal{D}$.

Hence $m-1 | n-1$.

(ii) $|\mathcal{D}| = b = (q^n - 1)(q^{n-1} - 1)/(q^m - 1)(q^{m-1} - 1)$. □

While no examples of nontrivial 2 – (n,m,1) designs are known, it has not been shown that no such designs exist for any pair of integers n,m for which $r,b \in \mathbb{N}$. My guess is that many such designs exist, but that their automorphism groups are extremely small. (This would account for the difficulty in finding examples. The usual method of constructing a design is to specify a small number of blocks $\{B_1,\ldots,B_s\}$, one from each orbit of the automorphism group Γ. Then $\mathcal{D} = \{B_i^\pi \mid 1 \le i < s, \pi \in \Gamma\}$).

Problem 4.3

Show that for each finite field $\mathbf{F} = GF(q)$, there exists an integer $N(q)$ such that: whenever $\mathcal{D} = (P,\mathcal{D})$ is a nontrivial 2-(n,m,1) design over $\mathbf{F}$ and $G = \mathrm{Aut}\mathcal{D}$, if $X \subseteq P$ satisfies $G_{\{X\}} / G_{(X)} \simeq \mathrm{Syn}(X)$, then $|X| < N(q)$.

Of course, this would eliminate case A. A special case of this conjecture can be deduced from Hering's theorem.

Theorem 4.4

Suppose that $\mathcal{D} = (P,\mathcal{D})$ is a nontrivial 2–(n,m,1) design over a finite field $\mathbf{F}$. If $G = \mathrm{Aut}\mathcal{D}$ acts transitively on P, then G is soluble.

Hering's theorem classifies the groups G which act transitively on the set of nonzero vectors of a finite vector space V. We can suppose that V is an n-dimensional vector space over the prime field GF(p), so that $G \leq GL(n,p)$. Let L be a subset of Hom(V,V) maximal with respect to the condition that L is normalized by G, L contains the identity and L is a field with respect to the addition and multiplication in Hom(V,V). (By Lemma 5.2 [9], L is unique unless $n = 2$, $p = 3$ and G is isomorphic to a quarterion group of order 8). There exist integers m and n^* such that $n = mn^*$, $|L| = p^m$ and n^* is the dimension of the vector space (V,L). Then $G \leq \Gamma L(V,L)$ must be one of the following types.

I. $SL(V,L) \leq G \leq \Gamma L(V,L)$.

II. There exists a nondegenerate skew-symmetric scalar product on (V,L) and G contains as a normal subgroup the group consisting of all isometries of the corresponding symplectic space.

III. $n^* = 6$, $p = 2$ and G contains a normal subgroup isomorphic to $G_2(2^m)$.

IV. There are finitely many exceptional groups. For our purposes, it is enough to know that in each of these cases $n = 2,4,6$. By Lemma 4.2, the corresponding projective spaces cannot carry nontrivial designs, and so these groups can safely be ignored.

This result also yields a classification of the groups $G \leq P\Gamma(n,q)$ which act transitively on the points of the n-dimensional projective space P over GF(q). Let $G^* \leq \Gamma L(n,q)$ be the preimage of G under the homomorphism $\Gamma L(n,q) \to P\Gamma L(n,q)$. Regard G^* as a subgroup of GL(N,p), where $q = p^t$ and $N = nt$. Let L be as above. Then GF(q) is a subfield of L, say $L = GF(q^r)$. If $r > 1$, then G preserves a geometric r-spread $\mathfrak{L}$ of P. This means that $\mathfrak{L}$ is a collection of r-subspaces which form a partition of P satisfying:

If $Q_1,Q_2,Q_3 \in \mathfrak{L}$ and $Q_3 \cap \langle Q_1,Q_2\rangle \neq \emptyset$, then $Q_3 \leq \langle Q_1,Q_2\rangle$. Define an incidence structure $P(\mathfrak{L})$as follows. The points are the elements of $\mathfrak{L}$ and the blocks are the sets of elements of $\mathfrak{L}$ contained in the subspaces $\langle Q_1,Q_2\rangle$ for $Q_1 \neq Q_2 \in \mathfrak{L}$. Then $P(\mathfrak{L})$ is an n/r-dimensional projective space over L. Furthermore, if G^* is of type II, then G preserves a symplectic polarity of $P(\mathfrak{L})$. A similar remark holds if G^* is of type III.

Proof of Theorem 4.4

First suppose that there exists r such that $n = rs$ and $PSL(s,q^r) \leq G \leq P\Gamma L(s,q^r)$. Clearly $s < n$, since otherwise G acts transitively on $P^{(m)}$, a contradiction. Hence G preserves a geometric r-spread $\mathfrak{L}$ for some $r \geq 2$. Suppose that $s > 1$. Let $X \in \mathfrak{L}$ and let $\ell \subseteq X$ be a line. Suppose that $\ell \subset Q \in \mathfrak{Q}$ and that $\gamma \in Q \setminus X$. Then the orbit of γ under $G_{(X)}$ contains at least $(q^n - 1/q^r - 1) - 1$ points. Hence

$$q^m - 1/q-1 = |Q| > q^n - 1/q^r - 1.$$

Since $r \leq n/2$ and $m \leq n+1/2$, this is impossible. Hence each $X \in \mathfrak{L}$ is a subdesign, and so $m\text{-}1 \mid r\text{-}1$. Let $X_1 \neq X_2 \in \mathfrak{L}$ and let $Y = \langle X_1,X_2 \rangle$. If Y is a subdesign, then $m\text{-}1 \mid 2r\text{-}1$. But then $m\text{-}1 \mid (r\text{-}1,2r\text{-}1) = 1$, a contradiction In particular, $s \geq 3$. Choose $\ell \subset Q \in \mathfrak{Q}$ such that $\ell \subset Y$ and $Q \not\subseteq Y$. Let $\gamma \in Q \setminus Y$. Then the orbit of γ under $G_{(Y)}$ contains at least q^{2r} points. Hence

$$q^{2r} < q^m\text{-}1/q\text{-}1 \leq q^r\text{-}1/q\text{-}1,$$

which is impossible. Hence $s = 1$ and G is soluble.

Next suppose that there exists $s \geq 4$ such that $PSp(s,q^r) \leq G$, where $n = rs$. Arguing as above, we find that $r > 1$ and that each $X \in \mathfrak{L}$ is a subdesign. Let $X_1,X_2 \in \mathfrak{L}$ be chosen so that $Y = \langle X_1,X_2 \rangle$ is a totally isotropic line of $P(\mathfrak{L})$. There exists a line $\ell \subset Y$ such that $Q \not\subseteq Y$, where $\ell \subset Q \in \mathfrak{Q}$. Let $\gamma \in Q \setminus Y$ and let $\gamma \in X_3 \in \mathfrak{L}$. Let $Z = \langle Y,X_3 \rangle$. Then Z is a plane of $P(\mathfrak{L})$ and $G_{(Y)}$ acts transitively on $\{X \in \mathfrak{L} \mid X \subseteq Z \setminus Y\}$. Hence the orbit of γ under $G_{(Y)}$ contains at least q^{2r} points, a contradiction.

Hence we can suppose that $q = 2^t$ and $G_2(2^{tr}) \leq G$, where $n = 6r$. By Lemma 4.2, $r > 1$. Let $X \in \mathfrak{L}$ and let $\ell \subset X$ be a line. Suppose that $Q \not\subseteq X$, where $\ell \subset Q \in \mathfrak{Q}$. Let $\gamma \in Q \setminus X$. By [7], the orbit of γ under $G_{(X)}$ contains at least $q^r(q^r + 1)$ points. Hence $q^m\text{-}1/q - 1 > q^r(q^r + 1)$, and so $m \leq 2r + 1 = n/3 + 1$, i.e., $n \leq 3(m - 1)$. Since $m - 1 \mid n - 1$, this implies that $n - 1 = 2(m - 1)$. But then $n = 2m - 1$ is odd, a contradiction. Hence each $X \in \mathcal{L}$ is a subdesign. Let $\mathcal{G}$ be the generalized hexagon of order (q^r, q^r) on $P(\mathcal{L})$ which is preserved by $G_2(q^r)$. Let $Y = \langle X_1, X_2 \rangle$ be a $\mathcal{G}$-line. Then we can suppose that there exists a line $\ell \subset Y$ such that $\ell \cap X_i \neq \emptyset$ for $i = 1,2$ and $Q \not\subseteq Y$, for $\ell \subset Q \in \mathcal{D}$. Let $\gamma \in Q \setminus Y$ and let $\gamma \in X_3 \in \mathcal{L}$. Since $G_2(q^r)$ acts transitively on the set of ordered ordinary hexagons of $\mathcal{G}$ [7], the orbit of γ under $G_{(X_1 \cup X_2)}$ contains at least q^{2r} points. Again, this is impossible. □

Corollary 4.5

Let $\mathcal{D}_3 = (P, \mathcal{D}_0)$ be the plane of $(M, \mathcal{D})$ generated by the Morley triple $\{\alpha, \beta, \gamma\}$. Then $H = \text{Aut}\, \mathcal{D}_3$ does not act transitively on P.

Proof

Arguing as in Lemma 3.2, we see that if $\langle \alpha, \beta \rangle \subset Q \in \mathcal{D}_0$, then $H_{\{Q\}}$ acts 2-transitively on Q. By [7], either $\text{Alt}(7) \leq H_{\{Q\}}/H_{(Q)}$ or $PSL(m,q) \leq H_{\{Q\}}/H_{(Q)}$. Since $m \geq 3$, H is not soluble. □

It seems likely that a more detailed study of $\mathcal{D}_3$ will allow us to eliminate case A. Let $\mathcal{L} = \{Q^{\pi} \mid \pi \in H\}$, where $\langle \alpha, \beta \rangle \subset Q \in \mathcal{D}_0$, and let $\Omega = \bigcup\{Q \mid Q \in \mathcal{L}\}$. The only candidates for the partial linear space $(\Omega, \mathcal{L})$ appear to be (possibly disjoint unions of) the flag-transitive generalized quadrangles, hexagons and octagons associated with various groups of Lie type. However, even these partial geometrics can be eliminated by looking at the actions of their automorphism groups on lines. I intend to say more on this matter in a later paper.

I end this section with an easy exercise. By Theorem 2.4 and the proof of Lemma 3.3, we already know that some of the planes of $(M, \mathcal{D})$ are neither projective nor affine. In fact, a stronger result holds.

Exercise 4.6

Prove that if $\mathcal{D}$ is a nontrivial 2-(n,m,1) design over a finite field, then $\mathcal{D}$ is neither a projective nor an affine plane.

5. EXTREMAL PROBLEMS

In this section, we consider projective space versions of theorems of Cameron [5],[6]. Let $G \leq P\Gamma L(\omega,q)$ have $n_k < \omega$ orbits on the set $PG^{(k)}(\omega,q)$ of k-subspaces of $PG(\omega,q)$ for each $k \in \mathbb{N}$. Fix an integer k such that $n_k > 1$, and let $\chi\colon PG^{(k)}(\omega,q) \to \{c_1,\ldots,c_{n_k}\}$ be a colouring which assigns distinct colours to the different G-orbits. The next result says roughly that G must locally preserve chains of subspaces. (Throughout, we adopt the convention that all r colours are used in an r-colouring).

Theorem 5.1 [18]

For all $r \geq 2$ and $t > k \geq 1$, there exists an integer $f(t,k,r)$ with the following property. Suppose that S is a projective space of GF(q) of dimension $m \geq f(t,k,r)$. If $S^{(k)}$ is r-coloured, then there exist

(i) subspaces $\emptyset \subseteq A \subsetneq B \subseteq C$ of S with dim $C = t$, dim $A = i$ and $\text{codim}_C\ B = j$, where $0 < i + j \leq k$, and

(ii) distinct colours c_1, c_2 such that $T \in C^{(k)}$ is coloured c_2 if $A \subseteq T$ and $\dim(T \cap B) = k - j$, and is coloured c_1 otherwise. □

We would like to use this local information to understand the global structure of the ω-categorical structure $M = \langle PG(\omega,q),\chi\rangle$. Unfortunately, very little can be said in general. However, in certain extremal circumstances, it is possible to use this local information to identify M.

If $L \in PG^{(k+1)}(\omega,q)$, then the <u>colour scheme</u> of L is the n_k-tuple $(a_1,\ldots,a_{n_k})$, where a_i is the number of elements of $L^{(k)}$ which have colour c_i. The <u>colour scheme matrix</u> A of the colouring χ is the matrix whose rows are the distinct colour schemes.

Theorem 5.2 [18]

The colour scheme matrix A has rank n_k. Furthermore, the colours and colour schemes can be ordered in such a way that the first n_k rows of A form a lower triangular matrix with nonzero diagonal entries. □

In particular, $n_k \le n_{k+1}$. Suppose that $n_k = n_{k+1}$. Then A is a nonsingular lower triangular matrix, and two (k+1)-subspaces lie in the same G-orbit if and only if they have the same colour scheme. Looking at the last row of A, we see that there exists a colour, say blue, which lies in a unique colour scheme. So after amalgamating the other colours and applying Theorem 5.1 to the resulting 2-colouring, we obtain

Theorem 5.3 [18]

Suppose that $G \le P\Gamma L(\omega,q)$ satisfies $1 < n_k = n_{k+1} < \omega$. Then there exists a G-invariant colouring φ: $PG^{(k)}(\omega,q) \to \{\text{red,blue}\}$ and integers i,j with $0 < i + j \le k$ such that

(i) G acts transitively on blue k-spaces and on (k+1)-spaces which contain a blue k-space; and

(ii) if $C \in PG^{(k+1)}(\omega,q)$ contains a blue k-space then there are subspaces $\varnothing \subseteq A \subsetneq B \subseteq C$ with $\dim A = i$ and $\mathrm{codim}_C B = j$ such that $T \in C^{(k)}$ is blue if and only if $A \subseteq T$ and $\dim(B \cap T) = k - j$. □

If k is small, this condition is strong enough to enable us to identify $M = \langle PG(\omega,q),\chi \rangle$. For example, the following is the main result of [18].

Theorem 5.4

Suppose that $q \ne 2$. If $G \le P\Gamma L(\omega,q)$ acts transitively on the points of $PG(\omega,q)$ and satisfies $1 < n_2 = n_3 < \omega$, then one of the following cases holds.

(i) G preserves a symplectic polarity of $PG(\omega,q)$ and acts transitively on the sets of totally isotropic lines and nonisotropic lines.

(ii) G preserves a geometric 2-spread L of PG(ω,q) and acts transitively on the planes of the incidence structure $P(L) \simeq PG(\omega,q^2)$. □

This is proved by working through the various possibilities for the pair of integers (i,j), $0 < i + j \leq 2$, given in Theorem 5.3. To give a flavor of the type of reasoning involved, I will give the details for two of the more interesting cases.

Suppose that $i = 1$ and $j = 0$. So if the plane Q contains a blue line, then there exists a point $\alpha(Q) \in Q$ such that $\ell \in Q^{(2)}$ is blue if and only if $\alpha(Q) \in \ell$. Let S be a finite dimensional subspace which contains a blue line, and let $\mathcal{B}_S = \{\ell \in S^{(2)} \mid \ell \text{ is blue}\}$. Suppose that $\ell \in \mathcal{B}_S$ and $\alpha \in S \setminus \ell$. Considering the plane $Q = <\ell,\alpha>$, we see that there is unique point $\beta \in \ell$ such that $<\alpha,\beta> \in \mathcal{B}_S$. Thus the incidence structure $(S,\mathcal{B}_S)$ ia a (possibly degenerate) generalized quadrangle. By Buekenhout and Lefevre [3], if dim $S > 6$ then there exists a point $\alpha \in S$ such that $\mathcal{B}_S = \{\ell \in S^{(2)} \mid \alpha \in \ell\}$. This implies that there is a point $\alpha \in PG(\omega,q)$ such that $\ell \in PG^{(2)}(\omega,q)$ is blue if and only if $\alpha \in \ell$, This contradicts the assumption that G acts transitively on points.

Suppose that $i = 0$ and $j = 2$. So if the plane Q contains a blue line, then there exists a point $\alpha(Q) \in Q$ such that $\ell \in Q^{(2)}$ is blue if and only if $\alpha(Q) \notin \ell$. Let S be a finite dimensional subspace, and consider the incidence structure $(S,\mathbb{R}_S)$, where $\mathbb{R}_S = \{\ell \in S^{(2)} \mid \ell \text{ is red}\}$. Let $\ell \in \mathbb{R}_S$ and $\alpha \in S \setminus \ell$. If the plane $Q = <\ell,\alpha>$ contains a blue line, then $\alpha(Q) \in \ell$ and $\alpha(Q)$ is the unique point of ℓ which is collinear with α in $(S,\mathbb{R}_S)$. Otherwise, α is collinear with every point of ℓ. By definition, $(S,\mathbb{R}_S)$ is a Shult space. By Buekenhout and Shult [4], there exists a (possibly degenerate) symplectic form σ on the underlying vector space of S such that $\mathbb{R}_S$ is the set of lines which are totally isotropic with respect to σ. It follows that there is a symplectic form σ on $V(\omega,q)$ such that $\ell \in PG^{(2)}(\omega,q)$ is red if and only if ℓ is totally isotropic with respect to σ. Since G acts transitively on points, σ is nondegenerate. Thus G preserves a symplectic polarity of PG(ω,q) and acts transitively on the set of nonisotropic lines. The reader is referred to [18] for a proof that G also acts transitively on totally isotropic lines.

If $G \leq P\Gamma L(\omega,q)$ is the full automorphism group of either a symplectic polarity or a geometric 2-spread of $PG(\omega,q)$, then G satisfies $n_{2k} = n_{2k+1} = k+1$ for each $k \in \mathbb{N}$.

Problem 5.5

Suppose that $G \leq P\Gamma L(\omega,q)$ acts transitively on the points of $PG(\omega,q)$ and satisfies $1 < n_k = n_{k+1} < \omega$ for some $k \in \mathbb{N}$. Does G preserve either a symplectic polarity or a geometric 2-spread of $PG(\omega,q)$?

This problem is open even for $k = 3$. In this case, the hardest situation to analyze is when each 4-subspace contains at most one blue plane. When this occurs, if $Q \in PG^{(3)}(\omega,q)$ is blue and $\ell \in Q^{(2)}$, then $acl(\ell) = Q$. A similar difficulty arises for larger values of k. Thus problem 5.5 provides further motivation for the problem of understanding expansions of $PG(\omega,q)$ in which certain subspaces are no longer algebraically closed.

REFERENCES

1. **R.H. Bruck**, A Survey of Binary Systems, Springer, 1966.
2. **F. Buekenhout**, Une caractérisation des espaces affins basée sur la notion de droite, Math. Z. 111 (1969), 367-371.
3. **F. Buekenhout** and **C. Lefèvre**, Generalized quadrangles in projective spaces, Arch. Math. 25 (1974), 540-552.
4. **F. Buekenhout** and **E.E. Shult**, On the foundations of polar geometry, Geo. Dedicata 3 (1974), 155-170.
5. **P.J. Cameron**, Orbits of permutation groups on unordered sets. J. London Math. Soc. (2) 17 (1978), 410-414.
6. **P.J. Cameron**, Orbits of permutation groups on unordered sets II, J. London Math. Soc. (2) 23 (1981), 249-264.
7. **P.J. Cameron** and **W.M. Kantor**, 2-transitive and antiflag transitive collineation groups of finite projective spaces, J. Algebra 60 (1979), 384-422.
8. **G.Cherlin, L. Harrington** and **A. Lachlan**, ω-categorical ω-stable structures, Annals of Pure and Applied Logic 28 (1985), 103-136.
9. **C.Hering**, Transitive linear groups and linear groups which contain irreducible subgroups of prime order, Geom. Dedicata 2 (1974), 425-460.
10. **C. Hering**, Transitive linear groups and linear groups which contain irreducible subgroups of prime order II, J. Algebra 93 (1985), 151-164.
11. **W.M. Kantor**, Homogeneous designs and geometric lattices, J. Combinatorial Theory, Series A, 38 (1985), 66-74.
12. **A. Lachlan**, Two conjectures regarding the stability of ω-categorical structures, Fund. Math. 81 (1973/74), 133-145.
13. **P.M. Neumann**, Some primitive permutation groups, Proc. London Math. Soc. 50 (1985), 265-281.
14. **M. O'Nan**, Automorphisms of unitary block designs, J. Algebra 20 (1972), 495-511.

15. **S. Shelah**, Classification theory and the number of non-isomorphic models, North-Holland, 1978.

16. **L. Teirlinck**, On linear spaces in which every plane is either projective or affine, Geom. Dedicata 4 (1975), 39-44.

17. **S. Thomas**, Designs over finite fields, Geom. Dedicata 24 (1987), 237-242.

18. **S. Thomas**, Groups acting on infinite dimensional projective spaces II, to appear in Proc. London Math. Soc.

19. **S. Thomas**, Triangle transitive Steiner triple systems, Geom. Dedicata 21 (1986), 19-20.

20. **H.P. Young**, Affine triple systems and matroid designs, Math. Z. 132 (1973), 343-359.

21. **K. Zsigmondy**, Zur Theorie des Potenzreste, Monatsh. für Math. u. Phys, 3 (1892), 265-284.

DEPARTMENT OF MATHEMATICS
RUTGERS UNIVERSITY
NEW BRUNSWICK
NEW JERSEY 08903

GENERIC FORMULAS AND TYPES A LA HODGES

by

Victor Harnik

University of Haifa

Mount Carmel, Haifa, Israel

§0. Introduction.

Let T be a complete theory with monster model $\mathbb{C}$ and A a subset of $\mathbb{C}$. Certain complete types $p \in S(\mathbb{C})$ have the "privilege" of being non forking over A. The smaller A is, the harder it is not to fork over it. Thus, the most "privileged" are those types that do not fork over the empty set $\emptyset$. If T is stable then, as we all know, non–forking types exist in sufficient abundance. If T happens also to have a group operation, then we are in the presence of a <u>stable group</u>.

The theory of stable groups has attracted much interest in recent years. There are several reasons for this interest. One is the fact that stable groups occur "in nature" more often than one might think. Another is that the theory of stable groups presents special features, due to the richer structure of the family of types. Indeed, the group itself acts on the family of types both from the left and from the right. If $p \in S(\mathbb{C})$ and g is an element of $\mathbb{C}$, then we define the left translate $gp \in S(\mathbb{C})$ of p by: $\varphi(x,\bar{a}) \in gp$ iff $\varphi(gx,\bar{a}) \in p$. In other words, gp is the type of any element of the form gc where c realizes the type p. The notion of right translate pg is defined analogously. Having these notions at hand, the following thought is quite natural: if $p \in S(\mathbb{C})$ does not fork over $\emptyset$, then it is a quite privileged type, but if it so happens that all its left translates also do not fork over $\emptyset$ then p is <u>truly</u> privileged. More formally, p is called a left–generic type iff gp does not fork over $\emptyset$ for all

$g \in \mathbb{C}$. The notion of right–generic type is defined similarly. These notions were introduced by Poizat in [8] (where he refers to earlier works of Zil'ber, Cherlin–Shelah [2] and his own [7], as a source of inspiration). Of course, these concepts would be uninteresting if it turned out that generic types are nonexistent or useless. Poizat proves their existence and illustrates their usefulness.

It turns out that generic types always exist in stable groups but there are very few of them. This last fact is illustrated by several results.

First, every left–generic type is also right–generic (cf. Poizat [8], Fait 5). In other words, we do not have two distinct families, of left– and right–generics, but one. Thus, we can speak simply of generic types (without "left/right" attribute).

A second result, pointed out by Hodges, is this. One might speak of a type p being generic over A, meaning that all its left– (and right–) translates do not fork over A. It turns out that if a type is generic over a small set A (i.e., $|A| < |\mathbb{C}|$) then it is generic (i.e., generic over $\varnothing$).

A third fact, due again to Poizat (Fait 6 in [8]) is that the generic types form precisely one orbit of the group action on $S(\mathbb{C})$. In other words, the group acts transitively on the family of generic types; stated in even simpler terms: if p is generic then q is generic iff $q = gp$ form some $g \in \mathbb{C}$ iff $q = ph$ for some $h \in \mathbb{C}$.

Poizat went on to "localize" the notion of genericity and defined the notion of generic formula (one that belongs to some generic type). He asked, in private conversations, for straightforward derivations of the properties of generic formulas (rather than inferring them from corresponding properties of generic types, as done in [8]). Hopefully, such proofs would be more elementary and simpler.

Wilfrid Hodges took up the challenge and found an elementary proof for a fundamental property of generic formulas (cf. 2.3 below and 1°, page 344 in [8]). From this, he gets simple proofs for the existence of generics and the equivalence of left– and right–genericity, two facts whose proofs by Poizat were heavy. We have thus an alternative approach to generic types. I lectured on Hodges' work in the Notre Dame Seminar on Stable Groups organized by A. Pillay in the Fall of 1986. This note, based on that lecture, appears with Hodges' kind approval. Also, several improvements upon a previous version are due to his useful comments (Hodges' expanded account of his work will appear in the second volume of a forthcoming book).

We should mention that another treatment dealing directly with generic formulas appears in Poizat's recently published book [9] (see esp. Section 5a). This is a new approach, more elegant than the one in [8]. It is less elementary than Hodges' method, at least in the technical sense of [3].

Still a different approach to generic types (but not formulas) is sketched in Hrushovski's [5], pp. 10–12.

I wish to thank the Mathematics Department of Notre Dame and especially J. Knight and A. Pillay for their hospitality.

§1. Preliminaries.

We use customary notations. T will be a complete stable theory with language L and monster model $\mathbb{C}$ (or **G** in case of groups, cf. §2 below). Models of T are always assumed to be elementary substructures of $\mathbb{C}$. a,b,... will be elements of $\mathbb{C}$, $\bar{a},\bar{b}$,... finite sequences of such elements, A,B,... small subsets of $\mathbb{C}$ and M,N,... models of T. L(A) will be L augmented with names for the elements of A. "$\models$" will denote satisfaction in $\mathbb{C}$.

Of the various known definitions of nonforking of formulas we adopt the following one (suggested by [1], [6], [10]).

Definition 1.1. $\varphi(x,\bar{c})$ does not fork over A iff it is almost satisfied in A, i.e., every model $M \supset A$ has an element satisfying the formula $\varphi(x,\bar{c})$.

By a standard compactness argument (as, e.g., 4.3 in [4]):

Lemma 1.2. $\varphi(x,\bar{c})$ does not fork over A iff for some $\delta(\bar{x}) \in L(A)$, $\bar{x} = (x_0,\ldots,x_{k-1})$,

$$\models \exists\, \bar{x}\, \delta(\bar{x}) \wedge \forall\, \bar{x}\, (\delta(\bar{x}) \rightarrow \bigvee_{i<k} \varphi(x_i, \bar{c})).$$

(We also say, if this is the case, that $\varphi(x,\bar{c})$ is almost satisfied over A via $\delta(\bar{x})$).

§2. Generic Formulas.

From now on we assume, unless otherwise stated, that the models of T have a group operation, i.e., T is the theory of a stable group. Accordingly, we let its monster model be **G** (rather than $\mathbb{C}$). The product of a,b $\in$ **G** will be ab.

<u>Remarks</u>. 1. Sometimes, the name "stable group" is associated with any stable structure in which one can define a group with definable universe or, even, with a universe that is defined by a set of formulas. The results presented here can be generalized to this situation (of course, most statements must be restricted to formulas or types that are consistent with G(x), where G is the formula, or set of formulas, defining the universe of the group).

2. Another direction of generalization was pointed out by W. Hodges: for most results, it is sufficient to assume that certain formulas (rather than the whole theory) are stable. As an example, Theorem 2.4 below is true whenever $\varphi(x,\bar{y})$ is a formula such that both formulas $\varphi'(x;v,\bar{y}) = \varphi(vx,\bar{y})$ and $\varphi''(x;v,\bar{y}) = \varphi(xv,\bar{y})$ are stable (such a formula φ is called by Hodges "bistable").

The crucial concept of this note is the following:

<u>Definition 2.1</u>: The formula $\varphi(x,\bar{c})$ is left–generic over A if for all $g \in \mathbf{G}$, $\varphi(gx,\bar{c})$ does not fork over A.

<u>Remark</u>. It is obvious that the extension of "left–generic" does not change if we replace in this definition, "for all $g \in \mathbf{G}$" by "for all $g \in M$ where M is any <u>saturated</u> model such that $A \cup \bar{c} \subset M$". It is less obvious that we can replace the same even by "for all $g \in M$ where M is <u>any</u> model such that $A \cup \bar{c} \subset M$". This follows from Poizat's [8].

<u>Lemma 2.2</u>. $\varphi(x,\bar{c})$ is left–generic over A iff there is $\delta(\bar{x}) \in L(A)$ such that

$$(*) \qquad \models \exists\bar{x}\, \delta(\bar{x}) \wedge \forall\bar{x}\,(\delta(\bar{x}) \rightarrow \forall v \bigvee_{i<k} \varphi(vx_i, \bar{c})).$$

Moreover, δ can be chosen to be left–invariant, i.e., for all $g \in \mathbf{G}$, $\models \delta(\bar{x}) \leftrightarrow \delta(g\bar{x})$ where $g\bar{x} = (gx_0,\ldots,gx_{k-1})$.

Proof (of the "only if" direction). If φ is left–generic over A then each g has a $\delta = \delta_g \in L(A)$ "witnessing" the nonforking of $\varphi(gx,\bar{c})$ in the sense of 1.2. It follows that the type

$$q(v) = \{\exists \bar{x}\,(\delta(\bar{x}) \wedge \neg \bigvee_{i<k} \varphi(vx_i, \bar{c})) : \delta \in L(A), \models \exists \bar{x}\, \delta(\bar{x})\}$$

is inconsistent and hence, there are $\delta_j(\bar{x}_j)$, $j < \ell$ such that for every $g \in \mathbf{G}$, some δ_j can serve as δ_g. Taking $\delta(\bar{x}_0, \bar{x}_1, \ldots, \bar{x}_{\ell-1}) = \bigwedge_{j<\ell} \delta_j(\bar{x}_j)$, we get (∗). To make δ left–invariant, replace it by $\delta'(\bar{x}) = \exists v\, \delta(v\bar{x})$. □

Remark. This lemma provides an elementary characterization of left–genericity. It is elementary in a technical sense (see e.g. [3]) because it can be stated as a Σ^0_2 formula in the language of second order arithmetic.

The following lemma is the main step in Hodges' treatment of generic formulas (compare with 1° page 344 of [8]). The proof adapts the argument of 4.4 in [3] to the present situation by making a clever use of property (*) below.

Lemma 2.3 ("Main Lemma"). If $\varphi(x,\bar{c}) \vee \psi(x,\bar{c})$ is left–generic over A then so is one of $\varphi(x,\bar{c})$, $\psi(x,\bar{c})$.

Proof. We are given that for some left–invariant $\delta(\bar{x}) \in L(A)$,

$$\models \exists \bar{x}\, \delta(\bar{x}) \wedge \forall \bar{x}\,(\delta(\bar{x}) \rightarrow \forall v \bigvee_{i<k} (\varphi(vx_i, \bar{c}) \vee \psi(vx_i, \bar{c}))).$$

Denote $\varphi^*(\bar{x}, v, \bar{y}) = \bigvee_{i<k} \varphi(vx_i, \bar{y})$. Notice that:

(*) $\models \varphi^*(\bar{x}; v, \bar{y}) \leftrightarrow \varphi^*(g\bar{x}; vg^{-1}, \bar{y})$.

Define ψ^* in a similar way. We then have:

(**) $\models \exists \bar{x}\, \delta(\bar{x}) \wedge \forall \bar{x}\,(\delta(\bar{x}) \rightarrow \forall v\,(\varphi^*(\bar{x}; v, \bar{c}) \vee \psi^*(\bar{x}; v, \bar{c})))$.

In this proof, by the φ^*–tree of height k defined by $\langle b_t, \bar{c}_t : t \in {}^{<k}2\rangle$ we shall mean the one whose branches are

$$\{\varphi^{*s(i)}(\bar{x}; b_{s|i}, \bar{c}_{s|i}) : i < k\},$$

for $s \in {}^k2$ (here $\varphi^{*0} = \varphi^*$ and $\varphi^{*1} = \neg\varphi^*$). Pictorially:

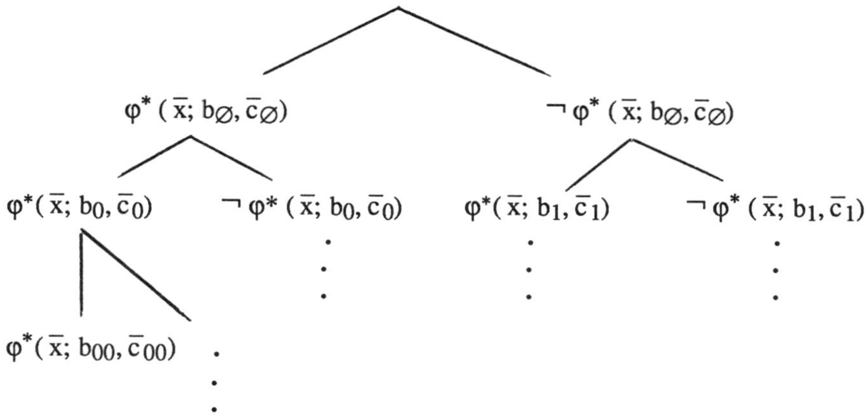

As the theory T is stable, so is the formula φ^* and hence, there is a finite upper bound on the heights of φ^*–trees with all branches consistent. It follows that there is a largest integer k such that there is a φ^*–tree of height k all of whose branches are consistent with $\delta(\bar{x})$. (This is the only place in this proof in which any stability assumption is used). Let $\lambda(<\bar{x}_s>, <v_t, \bar{y}_t>)$, where $<\bar{x}_s> = <\bar{x}_s: s \in {}^k2>$ and $<v_t, \bar{y}_t> = <v_t, \bar{y}_t; t \in {}^{<k}2>$, be the formula stating that, for all $s \in {}^k2$, $\bar{x}_s$ satisfies $\delta(\bar{x})$ as well as the s–branch of the φ^*–tree defined by $<v_t, \bar{y}_t>$.

By the left–invariance of $\delta(\bar{x})$ and by (*), we see that $\lambda(<\bar{x}_s>,<v_t, \bar{y}_t>) \leftrightarrow \lambda(<g\bar{x}_s>,<v_t g^{-1}, \bar{y}_t>)$.

It follows that if $\lambda^*(<\bar{x}_s>) = \exists<v_t, \bar{y}_t>\lambda$ then λ^* is left–invariant. Also, by the definition of k, we know that $\models \exists<\bar{x}_s>\lambda^*$.

<u>Claim</u>. Either $\models \forall <\bar{x}_s> \lambda^* \rightarrow \forall v \bigvee_{s\in {}^k2} \varphi^*(\bar{x}_s; v, \bar{c}))$ or $\models \forall <\bar{x}_s> \lambda^* \rightarrow \forall v \bigvee_{s\in {}^k2} \psi^*(\bar{x}_s; v, \bar{c}))$, hence either $\varphi(x, \bar{c})$ or $\psi(x, \bar{c})$ is left–generic over A.

<u>Proof of the Claim</u>: If not, then there are $<\bar{a}_s'>$, h such that $\models \lambda^*(<\bar{a}_s'>) \wedge \bigwedge_{s\in {}^k2} \neg\psi^*(\bar{a}_s'; h, \bar{c})$, which implies, by (**),

(i) $\models \lambda^*(<\bar{a}_s'>) \wedge \bigwedge_{s \in {}^k 2} \varphi^*(\bar{a}_s'; h, \bar{c})$

and, also, there are $\bar{a}_s''$, g such that

(ii) $\models \lambda^*(<\bar{a}_s''> \wedge \bigwedge_{s \in {}^k 2} \neg \varphi^*(\bar{a}_s''; g, \bar{c})$.

An examination of (i) and (ii) reveals a certain gap that is easily bridged: by (*) and the left invariance of λ^*, (i) implies:

(i)′ $\models \lambda^* (<g^{-1}h\, \bar{a}_s'>) \wedge \bigwedge_{s \in {}^k 2} \varphi^*(g^{-1}h\, \bar{a}_s'; g, \bar{c})$.

By the meaning of λ^*, (i)′ and (ii) imply the existence of φ^*–trees of height k defined by sequences $<g_t', \bar{c}_t'>$ and $<g_t'', \bar{c}_t''>$ whose branches are satisfied by $<g^{-1}h\, \bar{a}_s'>$ and $<\bar{a}_s''>$ respectively. (Keep in mind that these sequences satisfy $\delta(\bar{x})$ as well). We obtain a φ^*–tree of height k+1 defined by $<g_t, \bar{c}_t ; t \in {}^{<k+1}2>$ where $g_\varnothing = g$, $\bar{c}_\varnothing = \bar{c}$, $g_{0t} = g_t'$, $\bar{c}_{0t} = \bar{c}_t'$, $g_{1t} = g_t''$, $\bar{c}_{1t} = \bar{c}_t''$.

The branches of this tree are consistent with $\delta(\bar{x})$, a contradiction to the minimality of k. □

Let us remark that this proof is elementary in the sense that it can be formalized in RCA_0 (Recursive Comprehension Axiom with restricted induction, cf. [3]).

We now turn to corollaries of the Main Lemma. Notice first that 2.3 holds for right–genericity as well. This fact is immediately put to good use by Hodges:

Theorem 2.4. If $\varphi(x, \bar{c})$ is left–generic over A then it is right–generic over $\varnothing$. Hence, $\varphi(x, \bar{c})$ is left–generic over A iff it is right–generic over $\varnothing$ iff it is left–generic over $\varnothing$.

Proof. We are given that for some $\delta(\bar{x}) \in L(A)$,

$$\models \exists \bar{x}\ \delta(\bar{x}) \wedge \forall \bar{x} (\delta(\bar{x}) \rightarrow \bigvee_{i<k} \varphi(vx_i, \bar{c})).$$

Take $\bar{h} = <h_0,\ldots,h_{k-1}>$ such that $\models \delta(\bar{h})$. Then we have: $\models \forall v \bigvee_{i<k} \varphi(vh_i, \bar{c})$, hence the formula $\bigvee_{i<k} \varphi(vh_i, \bar{c})$ is right–generic over $\varnothing$. By 2.3 (applied to right–genericity) $\varphi(vh_i, \bar{c})$ is right–generic over $\varnothing$ for

some i. But then, for all $g \in \mathbf{G}$, $\varphi(vg, \bar{c}) \equiv \varphi(v(gh_i^{-1})h_i, \bar{c})$ does not fork over $\varnothing$. This means that $\varphi(v, \bar{c})$ or, if you wish, $\varphi(x, \bar{c})$ is right–generic over $\varnothing$. □

From now on, we say simply "generic" for "left– (or right)–generic over $\varnothing$".

If $X = \varphi(\mathbf{G}, \bar{c}) = \{a \in \mathbf{G} : \models \varphi[a, \bar{c}]\}$, then one denotes $gX = \{ga : a \in X\}$. The following easy corollary of 2.4 and its proof, shows the equivalence of 2.1 to Poizat's definition of a generic formula.

Theorem 2.5. Let $X = \varphi(\mathbf{G}, \bar{c})$. $\varphi(x, \bar{c})$ is generic iff there are $g_0,\dots,g_{k-1}$, such that $\mathbf{G} = g_0X \cup g_1X \cup \dots \cup g_{k-1}X$.

One more, quite straightforward, corollary of 2.3 is the existence of generic types (Theorem 3.1 below). The reader may go directly to the next section where we describe briefly this and a few other results on generic types; but if he selects to stay, we invite him to a discussion of the results presented so far.

Discussion. One can generalize 2.1 to formulas $\varphi(\bar{x}, \bar{c})$ with more than one free variable, by stipulating the nonforking of $\varphi(g\bar{x}, \bar{c})$ for all $g \in \mathbf{G}$. Lemmas 2.2 and 2.3 generalize to this context (with the same proof). However, 2.4 and 2.5 are not true anymore. Indeed, let $\varphi(x,y; c)$ be the formula $x^{-1}y = c$. This formula is left generic over $\{c\}$. If c is suitably chosen then $\varphi(x,y; c)$ is not left–generic over $\varnothing$ and not right–generic over $\{c\}$. If $c=1$ then $\varphi(x,y; c)$ is both left– and right–generic over $\varnothing$ but 2.5 does not hold for it.

This example is a particular case of a more general context. To present this context, let us return for a while to an arbitrary theory T with monster model $\mathbb{C}$. Assume nevertheless, that T has a definable group operation on a definable set G. Assume, furthermore, that we have a definable left–action of G on a definable set U. It is convenient to use a two sorted language with variables $v,u,v_1,u_1,\dots$ ranging over G and $x,x_1,\dots$ over U. Thus, a formula $\varphi(x, \bar{c})$ or a type $p(x)$ will always be assumed to imply $U(x)$.

In this general framework (which, by the way, has been considered also by Hrushovski in [5]) one can define left–generic formulas and prove 2.2 and 2.3. However, left–genericity over a set A does not imply left–genericity over $\varnothing$ as demonstrated by the formula $\varphi(x,y;c)$ discussed above (to fit that example into the general context, take $U = G \times G$ and define the action of G on U by $g(x,y) = (gx, gy)$). The same formula shows that 2.5 also fails.

Here is the place to ask a natural question. Returning for a moment to stable groups, is there a direct, transparent proof of the striking fact that left–genericity of a formula over a set implies left–genericity over $\varnothing$? Our discussion above shows that any such proof should use some special features that are not used in the proof of 2.3.

Back to our broader framework, we may ask ourselves where does the proof via 2.4 of the statement "left–generic over A implies left–genric over $\varnothing$" fail to generalize. The obvious obstacle is that we did not define what do we mean by right–genericity. One natural definition is the following. Say that $\varphi(x, \bar{c})$ is right–generic over A if for every $b \in U$, the formula $\varphi(vb, \bar{c})$ does not fork over A (of course, we assume that this formula implies $G(v)$). Lemma 2.2 holds for this notion and hence, right–genericity over a set A implies left–genericity over $\varnothing$. However, left–genericity does not imply right genericity. Still worse, the main Lemma 2.3 fails for right–generic formulas. Indeed, assume that we have an equivalence relation E on U with precisely two equivalence classes and such that G acts transitively on each class. Then, if $b,c \in U$ are representatives of the two classes then $E(x,b) \vee E(x,c)$ is right–generic over $\varnothing$ but neither $E(x,b)$ nor $E(x,c)$ are such (to see that such a situation can occur in a stable structure, take any stable group G, let $U = G \times \{b,c\}$ where b,c, are two distinct elements and for $(x,y), (x_1,y_1) \in U$ define $E((x,y),(x_1,y_1))$ iff $y=y_1$ and define the action of G on U by $g(x,y) = (gx,y)$). If we examine where does the proof of the main lemma fail, we are led to the conclusion that <u>if G happens to act transitively on U</u> then 2.3 does hold for right–genericity also. This much is easy to verify. Under the same assumption (that G acts transitively on U), 2.4 and hence, 2.5 are true as well.

To see that 2.4 holds indeed, we need a general fact which we state using the terminology introduced in Lemma 1.2 (and assuming that $u, v, v_1, \ldots$ range over a group G definable in a stable theory).

Theorem 2.6. If $\varphi(v, \bar{c}) \vee \psi(v, \bar{c})$ is almost satisfied over A via a left–(right)–invariant formula δ then one of φ, ψ is almost satisfied over A via a left–(right) invariant formula λ.

Sketch of the proof: the assumption implies that

$$\models \exists \bar{v}\, \delta(\bar{v}) \wedge \forall \bar{v}\, (\delta(\bar{v}) \rightarrow \forall u \bigvee_{i<k} (\varphi(uv_i, \bar{c}) \vee \psi(uv_i, \bar{c}))).$$

From this point on, proceed as in the proof of 2.3. □

Returning to the generalized 2.4, if $\varphi(x, \bar{c})$ is left–generic over A then, as in the proof of the original 2.4, we conclude that $\models \forall v \bigvee_{i<k} \varphi(vh_i, \bar{c})$ for a suitable $\bar{h}$. By 2.6, there is $i<k$ such that $\varphi(vh_i, \bar{c})$ is almost satisfied over $\varnothing$ via a right–invariant formula λ. If G acts transitively on U this implies immediately the right–genericity of $\varphi(x, \bar{c})$.

Remark. The assumption that G acts transitively on U is also needed by Hrushovski in [5] in order to get a smooth theory of generic types.

Another natural variant of right–genericity is the following. Assume that G also acts on U from the right in a definable way. In this case right–genericity has an obvious definition and 2.3 holds for both notions of one–sided genericity. However, 2.4 and 2.5 fail even if we assume the natural assumption of associativity: $(ga)h = g(ah)$ for all $g,h \in G$, $a \in U$. But again, if we assume in addition that G acts transitively on U from both the left and the right then 2.4 and 2.5 are true as well.

§3. Generic types.

We return to stable groups.

Lemma 3.1. If Γ is a (not necessarily complete) type closed under finite conjunctions all of whose formulas are generic then there is a type $p \in S(G)$ such that $p \supset \Gamma$ and all formulas of p are generic.

The proof uses 2.3 and is a standard application of Zorn's Lemma.

This result motivates the following:

Definition 3.2. A type p is generic iff all its formulas are generic.

This definition, in which p is supposed to be a complete type over any set, transforms 3.1 into Poizat's theorem on the existence of generic types. Some of the other results of [8] follow quite easily. Thus if p is generic and $q \supset p$ then q is generic iff q is a nonforking extension of p ("only if" is trivial while "if" follows using a result of Lascar stating that nonforking extensions of a given type $p \in S(A)$ can be mapped onto each other by isomorphisms over A – cf. e.g. 5.1(i) in [4]). Another, quite easily seen result is that for $p \in S(\mathbf{G})$, p is generic iff for all $g \in \mathbf{G}$, gp does not fork over $\varnothing$. Also, if $p \in S(\mathbf{G})$ is generic then so is p^{-1}, the type of any element of the form c^{-1} where c realizes p. These and other remarkable results can be found in [8]. One result that does not appear there is due to Hodges as we mentioned in the introduction:

Theorem 3.3. If $p \in S(\mathbf{G})$ is left– (or right–) generic over a small set A (meaning that, for all $g \in \mathbf{G}$, gp does not fork over A) then it is generic.

If $p \in S(\mathbf{G})$ is generic then so is gp for all $g \in \mathbf{G}$. Thus g acts on the family of generic types. We close with a proof of Poizat's result, also mentioned in the introduction, that this action is transitive:

Theorem 3.4. If $p_1, p_2 \in S(\mathbf{G})$ are generic types then for some $g \in \mathbf{G}$, $gp_1 = p_2$.

Proof. If $\varphi(x) \in p_2$ then $\varphi(x)$ is generic and by 2.5, there are $h_0,\dots,h_{k-1}$ such that $\models \forall x \bigvee_{i<k} \varphi(h_i x)$.

Thus, for some h, $\varphi(hx)$ is consistent with p_1, hence, belongs to p_1. For $\varphi(x, \bar{y}) \in L$, let $\varphi'(x; v, \bar{y}) = \varphi(vx, \bar{y})$ and let $\theta_\varphi(v, \bar{y})$ be a φ'–definition of p_1. Take a small $M \prec \mathbf{G}$. The type

$$q(v) = \{\theta_\varphi(v, \bar{c}) \colon \varphi(x, \bar{c}) \in p_2 \restriction M\}$$

is consistent, by the opening remark of this proof. Let g realize q. Then

$$\varphi(x) \in p_2 \restriction M \Rightarrow \varphi(gx) \in p_1 \Rightarrow \varphi(x) \in gp_1.$$

It follows that $gp_1 \restriction M = p_2 \restriction M$ and as both gp_1 and p_2 do not fork over M, we have $gp_1 = p_2$. □

REFERENCES

[1] J. Baldwin, Strong saturation and the foundations of stability theory, Logic Colloquium 82, G. Lolli et al. eds., North–Holland 1984, 71–84.

[2] G. Cherlin and S. Shelah, Superstable fields and groups, Annals of Math. Logic 18 (1980), 227–270.

[3] V. Harnik, Stability theory and set existence axioms, Journal of Symbolic Logic 50 (1985), 123–137.

[4] V. Harnik and L. Harrington, Fundamentals of forking, Annals of Pure and Applied Logic 26 (1984), 245–286.

[5] E. Hrushovski, Contributions to stable model theory, Ph.D. Dissertation, University of Calif. at Berkeley, 1986.

[6] A. Pillay, Forking, normalization and canonical bases, Annals of Pure and Applied Logic 32 (1986), 61–81.

[7] B. Poizat, Sous–groupes définissables d'un groupe stable, Journal of Symbolic Logic 46 (1981), 137–146.

[8] B. Poizat, Groupes stables avec type génériques réguliers, Journal of Symbolic Logic 48 (1983), 339–355.

[9] B. Poizat, Groupes Stables, Nur al–Mantiq wal–Ma'rifah, Villeurbanne, 1987.

[10] P. Rothmaler, Another treatment of the foundations of forking theory, Proceedings of the First Easter Conference on Model Theory, Diedrichsgaben DDR, Humboldt–University 1983.

SOME NOTES ON STABLE GROUPS

John T. Baldwin*

Department of Mathematics, Statistics and Computer Science

University of Illinois at Chicago

March 24, 1988

The best examples of ω–stable groups are algebraic groups over the complex numbers. We try in Section 1 of this exposition to clarify what some model theoretic concepts mean in some quite concrete situations and write down for reference several simple facts which several model theorists have reconstructed several times each. We raise further questions directed at extensions of the Cherlin conjecture. These questions and related remarks are aimed at clarifying the distinction between the group theoretic and the geometric properties of an algebraic group. They emphasize the oft–mentioned, in the abstract, insistence that a stable group may have further structure. The comments here arise from lectures given at Notre Dame during the 1986-87 year. Several of the results arose in a number of discussions with among others Cherlin, Loveys, MacPherson, Marker, Martin, Nesin, Pillay, Steinhorn and Tanaka.

---

* Partially supported by NSF grant DMS 8602588; much of the work on this paper was done during the author's visit to Notre Dame.

In Section 2 we explore the difference between the use of the term 'automorphism group' in algebraic geometry and group theory. We show in Section 3 that although earlier examples have shown that the stability spectrum is not preserved by taking a finite extension of a group, every extension of Z^n by a finite group is superstable. In other words all crystallographic groups are superstable.

1. ω–STABLE AND ALGEBRAIC GROUPS

It is well known that any matrix group which is definable over the complex numbers is ω–stable of finite rank; indeed these provide the main examples of such groups. Cherlin conjectured that a simple ω–stable group of finite rank is an algebraic group over an algebraically closed field. We will discuss the meaning of 'is' in this context and some extensions of the conjecture to groups which are not simple. We proceed primarily by studying a few examples.

Since the class of ω–stable groups is closed under product, taking the product of algebraic groups of different characteristic we obtain ω–stable groups which are not algebraic. Another example is the group Z_{p^∞}. Our main interest here is with a distinction which arises already when considering algebraic groups.

There are two natural ways for a model theorist to view an algebraic subgroup of GL(n,$\mathbb{C}$) (the n x n invertible matrices over the complex numbers): as a *pure* group (G,·) or as an algebraic group G* = (G,·, R_i) where the R_i are the restrictions to G of all relations on $\mathbb{C}^{n^2}$ definable (without parameters) in the structure ($\mathbb{C}$, +,·). We refer to the language of the second structure as the *geometric language*. We discuss below which concepts concerning algebraic groups are defined in the group language and which in the geometric. But sometimes there is no difference. The following question was essentially posed by Poizat.

1 Question: *For which affine algebraic groups are* (G,·) and G* biinterpretable?

It is natural to view this question over an arbitrary algebraically closed field. In both this question and the next we have restricted for concreteness to affine algebraic groups (where the universe of the group is a Zariski closed subgrtoup of $\mathbb{C}^{n^2}$). The questions are equally meaningful for arbitrary algebraic groups (since an abstract variety can be viewed as finite union of affine varieties modulo a definable equivalence relation). By Zil'ber, Poizat, van den Dries, and Hrushovski the structures G and G* are biinterpretable if G is simple (Theorem 4.16 of [17]). Since the projection functions are definable, the structure G* is always ω_1– categorical. So ω_1– categoricity is a necessary but not sufficient condition for a positive answer to Question 1. It is not sufficient since G $= (\mathbb{Z}_{p^2})^{\omega}$ is ω_1– categorical and (as noted by Nesin) can be viewed as an algebraic group as follows. G is elementarily equivalent to the algebraic group over an infinite field k of characteristic 2

$$\left\{\begin{pmatrix}1 & a & b\\ 0 & 1 & a\\ 0 & 0 & 1\end{pmatrix} : a,b \in k\right\}.$$

With a little more background we will see that in this case (G,·) is not biinterpretable with G*. Recall that a (necessarily ω_1–categorical) structure A is *almost strongly minimal* if, possibly expanding the language to name a finite number of elements that realize a principal type, there is a definable subset D of A which is strongly minimal such that $A \subseteq \text{acl } D$. But $(\mathbb{Z}_{p^2})^{\omega}$ is not almost strongly minimal and, again, the existence of coordinate functions implies that G* is always almost strongly minimal.

2 Question: *Characterize the* ω_1*–categorical affine algebraic groups.*

It is hard to imagine a purely algebraic classification since there are examples of ω_1–categorical algebraic groups which are simple, solvable but not nilpotent, and nilpotent. Every nonabelian Morley rank 2 group is ω_1–categorical. This was shown by Cherlin [4] in the solvable but not nilpotent case, by Tanaka [18] in the nilpotent case, and finally by a general argument of Lascar [12]. Our use of almost strong minimality to show the geometry is not defined in $(\mathbb{Z}_{p^2})^\omega$ leads to the following refinement of the question.

3 Question: *Which ω_1–categorical groups are almost strongly minimal?*

We will give examples below of algebraic groups in which the field is definable but which are not ω_1–categorical. Thus the program for proving some ω–stable group G of finite Morley rank is an algebraic group has two distinct steps:

i) Define a field k, necessarily algebraically closed, in G.
ii) Show G is an algebraic group over k.

The second step remains the sticking point for rank 2 nilpotent groups. Tanaka [18] has extended the analysis in [4] to show every such group is ω_1–categorical. All the known examples (cf. [18] and [13]) are in fact algebraic groups. Nesin has shown ([13]) how to define the field in the group language. The following observation was made by James Loveys and myself.

1.1 Lemma. *If* G *is a Morley rank 2 nilpotent group then* G *is not almost strongly minimal.*

Proof. Let A denote $(\mathbb{Z}_p)^{\omega_1}$. Then any group in our class (up to elementary equivalence) may be represented as a central extension of A by itself [4]. That is, we can represent G as A x A with the multiplication given by (a, b) (c, d) = (a + c, b + d + f(a, c)) where f : A x A $\rightarrow$ A is a cocycle satisfying certain equations (specified in e.g. [18]). Now let α be a group homomorphism of A into A. Then it is easy to check that $\hat{\alpha}$ defined by

$\hat{\alpha}(a, b) = (a, b + \alpha(a))$ is an automorphism of G which fixes the center of G pointwise. To see G is not almost strongly minimal note that for any $X = \{(a_1,b_1),\ldots,(a_n,b_n)\}$ and any (a, b) with a not in the (finite) subgroup generated by the a_i, it is possible to choose infinitely many homomorphisms α of A which map $a_1,\ldots, a_n$ to 0 and differ on a. Thus the associated $\hat{\alpha}$ demonstrate that $(a, b) \notin \operatorname{acl}(X \cup Z(G))$. If G is almost strongly minimal, for any infinite definable set W there is a finite set X with $G \subseteq \operatorname{acl}(X \cup W)$. With this contradiction we finish. □

We would like to classify Morley rank 2 nilpotent nonabelian groups. All known examples are definable in algebraically closed fields. As noted before, Nesin has shown the field is definable in basic example. Is a field definable in every such example? The preceding result shows that even for the examples which are algebraic groups it is impossible to define the full geometric structure in the group language. The question which remains is whether any cocycle f which gives a rank 2 nilpotent group must be definable in the field structure.

We will explore the relations between these questions by considering some subgroups of the 2 x 2 matrices over the complex numbers. In describing these examples we not only give concrete examples but report some algebraic folklore which is helpful in the model theoretic context. We call a matrix *diagonal* if all entries off the main diagonal are zero and *scalar* if it is a diagonal matrix with all its nonzero entries equal. We fix the following notation.

$$G = GL(2,\mathbb{C}) = \left\{\begin{pmatrix} a & b \\ c & d \end{pmatrix} : ad - bc \neq 0\right\} \quad \text{general linear}$$

$$S = SL(2,\mathbb{C}) = \left\{\begin{pmatrix} a & b \\ c & d \end{pmatrix} : ad - bc = 1\right\} \quad \text{special linear}$$

$$Z = Z(2,\mathbb{C}) = \left\{\begin{pmatrix} a & 0 \\ 0 & a \end{pmatrix} : a \neq 0\right\} \quad \text{center of G}$$

$$Z_0 = Z_0(2,\mathbb{C}) = \left\{\begin{pmatrix} a & 0 \\ 0 & a \end{pmatrix} : a^2 = 1\right\} \quad \text{center of S}$$

$PSL(2,\mathbb{C}) = S/Z_0$ projective special linear

$PGL(2,\mathbb{C}) = G/Z$ projective general linear

Note that $PGL(2,\mathbb{C})$ and $PSL(2,\mathbb{C})$ are isomorphic groups. For, PSL is defined to be $S/Z \cap S \approx Z \cdot S/Z$. But since $\mathbb{C}$ is algebraically closed each matrix A can be factored as a product of a scalar matrix and one of determinant one. So $Z \cdot S = G$. Of course these two groups are quite different over finite fields. The word 'projective' arises because the action of G on affine space becomes a faithful action on projective space when the center is factored out. The special linear group is definable in the general linear group; it is the commutator subgroup. In general the commutator subgroup is not definable. However, it is in this case. It is well known to algebraists that there is a bound on the number of multiplications needed to generate the commutator subgroup of an algebraic group over $\mathbb{C}$; a consequence of the Zil'ber indecomposability theorem extends this to any connected ω–stable group of finite Morley rank.

In the model theoretic context we say a (definable) subgroup is connected if it has no definable subgroup of finite index (equivalently has Morley degree 1). For any subset H of an ω-stable group let $\tilde{H}$ be the minimal definable subgroup containing $\tilde{H}$. We occasionally use this notation below in the context of stable groups. In such cases we implicitly assume that the group is saturated and $\tilde{H}$ is the minimal type–definable subgroup containg H (see[12]).

1.2 Definition. A *Borel* subgroup of an algebraic group G is a maximal connected solvable subgroup of G.

1.3 Fact. Every Borel subgroup of an algebraic group (over an algebraically closed field) is definable in the pure group language.

Proof. Recall that it was shown by Zil'ber [19] for ω–stable groups of finite rank and by [2] for stable groups that if H is solvable then so is $\tilde{H}$. So a subgroup of a stable group which is maximal among the solvable subgroups definable in the geometric language is definable in the group language. Now if G is an algebraic group and B is Borel, then B is maximal among all solvable closed subgroups not just the connected ones (See [11] Corollary A of Section 23) so B is definable as required. □

There may exist other maximal closed solvable groups. They will be definable but have lower rank than a Borel. There are examples of finite maximal solvable subgroups of semisimple algebraic groups (cf. [16]).

Here is a natural representation for a Borel subgroup of GL(2,$\mathbb{C}$).

$$B = B(2,\mathbb{C}) = \left\{\begin{pmatrix} a & b \\ 0 & d \end{pmatrix} : ad \neq 0\right\} \quad \text{Borel subgroup}$$

Such a representation can be explained as follows. For any n, it is easy to see that the group of upper triangular matrices is solvable. (The first commutator subgroup has ones on the diagonal, the second has zeros on the superdiagonal, the third has zeros on the superdiagonal and on the second superdiagonal, etc.) The Lie–Kolchin–Malcev theorem asserts that every solvable subgroup is conjugate to a subgroup of upper triangular matrices. Thus the group of upper triangular matrices is a Borel subgroup.

The argument for the solvability of the upper triangular matrices shows that the commutator subgroup of the upper triangular matrices is nilpotent. Nesin [15] has extended this result by showing that the commutator subgroup of a connected solvable group of finite Morley rank is nilpotent.

In our further discussion we will need two more subgroups.

$$U = U(2,\mathbb{C}) = \left\{\begin{pmatrix} 1 & b \\ 0 & 1 \end{pmatrix} : b \in \mathbb{C}\right\} \quad \text{unipotent}$$

$$T = T(2,\mathbb{C}) = \left\{\begin{pmatrix} a & 0 \\ 0 & d \end{pmatrix} : ad \neq 0\right\} \quad \text{torus}$$

The properties defining these groups are expressible in the *geometric* language or more precisely in the language of matrix rings. An element a of a matrix group is called *unipotent* if for some n, $(a-1)^n = 0$, a closed subgroup is unipotent if all its elements are. Note that an upper triangular matrix with ones on the diagonal is unipotent. The subgroup U is a minimal definable infinite unipotent group. Note that $T(n,\mathbb{C})$ is abstractly isomorphic to a direct product of n copies of the multiplicative group of the field. That is, as a subgroup of $GL(n,\mathbb{C})$, its elements are simultaneously diagonalizable. As a maximal Abelian subgroup of a stable group, T is definable with parameters in G ($T = Z(C_G(T_0))$ for a finite subset T_0 of T).

1.4 Lemma. *Consider the pure group* G. *Suppose* $G = G_0 \times G_1$ *where both* G_0 and G_1 *are infinite and* G_0 is *∅–definable in G. Then* G *is not* ω_1*–categorical.*

Proof. We can apply the Feferman–Vaught theorem to obtain a two cardinal model of Th(G). □

For any group H and finite subgroup F, if H is ω_1–categorical then so is H/F. The converse is false; an example of an unstable group G with a finite center Z such that G/Z is ω_1–categorical is given in [14]. But this does not stop us from proving the next result.

1.5 Fact. $GL(2,\mathbb{C})$ is not ω_1–categorical.

Proof. Note that

$$GL(2,\mathbb{C}) \approx (SL(2,\mathbb{C}) \times Z)/\hat{Z}_0$$

where $\hat{Z}_0$ is the set of pairs (a,a) contained in SL(2,$\mathbb{C}$) x Z with $a \in Z_0$. As noted in the last lemma SL(2,$\mathbb{C}$) x Z has a two cardinal model and this property is preserved by factoring out a finite normal subgroup; so GL(2,$\mathbb{C}$) is not ω_1–categorical. □

So GL(2,$\mathbb{C}$) is another example of an algebraic group where the geometric language is more expressive than the group theoretic. Now the same argument shows

1.6 Fact. B(2,$\mathbb{C}$) is not ω_1–categorical.

But this fact provides an interesting anomaly. It is easy to see that (T,·) = T(2,$\mathbb{C}$) is strongly minimal and so ω_1–categorical. But it is easy to verify

1.7 Proposition. $B \subseteq \mathrm{dcl}\,(T \cup \{\begin{pmatrix} 1 & 1 \\ 0 & 1 \end{pmatrix}\})$.

Proof. $\begin{pmatrix} a & b \\ 0 & c \end{pmatrix} = \begin{pmatrix} 1 & 0 \\ 0 & cb^{-1} \end{pmatrix}\begin{pmatrix} 1 & 1 \\ 0 & 1 \end{pmatrix}\begin{pmatrix} a & 0 \\ 0 & b \end{pmatrix}$ if $b \neq 0$. If $b = 0$, there is nothing to prove. □

Why can't we conclude that B is ω_1–categorical? To explore this question we introduce some further terminology.

1.8 Definition. Let G_0 be a definable substructure of an L–structure G. We denote by $G|G_0$ the structure with universe G_0 and as n–ary relations all restrictions to G_0^n of Ø–definable relations on G^n. Then we say G_0 is *full* in G if the L structure G_0 is interdefinable with $G|G_0$.

Since (B,·) is not ω_1–categorical but (T,·) is, we infer that T is not full in B. Just what structure does B impose on T ? We answer this question by describing a structure on T which is definable in B and such that every automorphism α of this structure extends to an automorphism of B. Note that T is definable in the group language of B (with parameters) but is not Ø–definable in B.

To make this description note first that the scalar matrices (Z) of T, although not definable in (T,·) are definable in (B,·) as the center of B. Now the map

$$\lambda: \begin{pmatrix} a & 0 \\ 0 & b \end{pmatrix} \mapsto \begin{pmatrix} a & 0 \\ 0 & b \end{pmatrix}\begin{pmatrix} 1 & 1 \\ 0 & 1 \end{pmatrix}\begin{pmatrix} a^{-1} & 0 \\ 0 & b^{-1} \end{pmatrix} = \begin{pmatrix} 1 & ab^{-1} \\ 0 & 1 \end{pmatrix}$$

identifies T/Z with U minus the identity element. Define an addition on T/Z by setting $A \oplus B = \gamma^{-1}(\gamma(A) \cdot \gamma(B))$. A straightforward computation verifies that $(T/Z, \oplus, \cdot)$ is a field and of course γ is an isomorphism between$(T/Z, \oplus)$ and (U,·). We claim the structure B imposes on T is interdefinable with the structure obtained by expanding the language to name the center and adding $\oplus$ as an operation on T/Z. To see this we show that any automorphism α of the group structure on T which induces an automorphism of the field structure on T/Z extends to an automorphism of (B, ·). Let θ denote the action of T on U by conjugation. Then

$$B \approx U \rtimes_\theta T$$

and identifying U with T/Z via γ

$$B \approx T/Z \rtimes_\theta T.$$

Now extend α to $\hat{\alpha} : B \to B$ by $\hat{\alpha}(u,t) = (\alpha u, \alpha t)$. Direct computation verifies that $\hat{\alpha}$ is an automorphism of B.

Summing up,

1.9 Lemma. *As a subgroup of the Borel subgroup B, the torus T has a distinguished subgroup Z and T/Z is an algebraically closed field; T has no further structure.*

As a counterpoint to this example consider the Borel subgroup B′ of SL(2, ℂ): $\left\{\begin{pmatrix} a & b \\ 0 & d \end{pmatrix} : ad = 1\right\}$. It is easy to see B′ is in the algebraic closure of the upper triangular subgroup U′. To see it is in the definable closure is harder. Define a 1–1 map from U′ to the diagonal elements by sending $u \in U'$ to the unique diagonal t such that for some diagonal x, $x^2 = t$ and x conjugates $\begin{pmatrix} 1 & 1 \\ 0 & 1 \end{pmatrix}$ to u.

2 THE AUTOMORPHISM PUN

Let the multiplicative group k^x act by multiplication on the additive group k^+. It is a well known geometric fact that the 'automorphism group' of the affine line over k is isomorphic to the semidirect product of k^+ by k^x. Each element can be thought of as a pair (a,b) which acts on k by sending x to ax + b. In this section we will call this group the affine group. By 'automorphism group', the geometer does not mean the group of permutations which preserve a certain set of relations on the line. Rather he means the collection of 1 – 1 self morphisms (i.e. polynomial maps) of the line to itself whose inverses are also morphisms. The Frobenius map is the standard example that this is not the same as a bijective morphism. Note however, that a bijective definable map has a definable inverse.

4 Question: *Find a natural (in particular finite) set of relations* $\bar{R}$ *on the set of complex numbers such that the set of automorphisms of* $(\mathbb{C}, \bar{R})$ *is isomorphic to the affine group.*

Note that if the relations chosen are definable from the field structure, any field automorphism (in the usual sense) would induce an automorphism of the structure. So if we permit arbitrary abstract automorphisms the relations imposed must include some which are not definable from the field structure. Since the affine group acts strictly 2–transitively on $\mathbb{C}$, it suffices to name each orbit of three tuples to solve the question. But this language is uncountable and no countable sublanguage of it will suffice.

A more tractable problem is to find a structure with universe $\mathbb{C}$ whose group of *definable* automorphisms is isomorphic to the affine group. With Pillay, Steinhorn and Loveys we arrived at the following solution. Put on $\mathbb{C}$ the operations

$$\oplus(x,y,z) = x + y - z$$

and

$$\otimes(x,y,z,w) = ((x-z)(y-z)/(w-z)) + z.$$

That is, hide both the zero and the one.

Now naming 0 and 1 allows one to recover the field and it admits elimination of quantifiers. Suppose that $\varphi(x,y)$ defines a permutation α of k. Then in characteristic zero we will show φ is equivalent to a formula of the form $f(x,y) = 0$ where f is a linear polynomial. It has been proved (e.g. [17] Chapter 4.c) that a definable function defined on a variety is rational on a Zariski open subset of the variety.

Since we are in characteristic 0 (and the field is algebraically closed) a rational function is $1-1$ with infinite domain only if it is the ratio of linear maps. Thus φ defines the graph of a fractional linear transformation. A routine computation shows the only fractional linear transformations which preserve the relation $\oplus(x,y,z) = w$ are of the form $ax + b$. Now $(k,+)$ is irreducible as an ω–stable group. Thus every element is a sum of generics and so since α preserves $\oplus$ its definition as a linear function on an open set extends to all of k and we finish.

The structure described above is inadequate to solve the problem in characteristic p. In fact, no set of relations on $\mathbb{C}$ which are definable from the field structure and from which multiplication is defined can give the affine group as the group of definable automorphisms. For, the Frobenius automorphism would be definable and not in the affine group.

Given any set A and group of permutations G of A, the canonical language associated with G and A has as relations the subsets of A^n which are left invariant by the action of G. Lemma 1.9 and our last remark lead to the following question.

5 Question: *When does the canonical language for A and G have a finite basis (i.e.* G = Aut $(A,R_1,...,R_n)$?

Hrushovski [10] has shown that assuming the structure on A is ω–stable and ω–categorical the language can be taken finite; can anything reasonable be said in more generality? In [9], the proof of Representation Lemma 2 on page 63 shows that if the action of G is definable then there is a bound on the arity of relations which must be added to form a basis.

Another puzzle is to find a set of relations on $\mathbb{C} \cup \{\infty\}$ to make the definable automorphism group of the resulting structure be the fractional linear transformations. That is, to solve for the projective line the problem we have solved here for the affine line. Hrushovski has indicated informally a solution to this problem and its generalization to an arbitrary curve.

3 CRYSTALLOGRAPHIC GROUPS

In this section we show that a class of groups which have been studied intensively by both mathematicians and chemists are superstable of finite U–rank. Along the way we recall some examples of Cherlin and Rosenstein and Thomas which show the result is a little more special than one might hope. All of these are concerned with the preservation of the stability classification by finite extensions of groups.

For our purposes we take as a definition the conclusion to a theorem of Zassenhaus giving the following algebraic equivalent of the standard definition of a crystallographic group as a group of symmetries of real n–space [8].

3.1 Definition. G is a crystallographic group if G has a maximal abelian subgroup A such that $A \triangleleft G$, G/A is finite and A is isomorphic to $\mathbb{Z}^n$ for some n.

The first step in the argument that every crystallographic group is superstable is to recall from [3] that every abelian by finite group is stable. But taking finite extensions can disturb the stability spectrum. The following example was pointed out to us by Simon Thomas. Such an example was requested in [14].

3.2 Example. Let $G_1 = GL(2,\mathbb{C})$. Embed $\mathbb{Z}_2 = (\sigma)$ into the group of automorphisms of G_1 by setting

$$\sigma\left(\begin{pmatrix} a & b \\ c & d \end{pmatrix}\right) = \begin{pmatrix} \bar{a} & \bar{b} \\ \bar{c} & \bar{d} \end{pmatrix}$$

where $\bar{a}$ denotes the complex conjugate of a. Now if G_2 is the semidirect product of G_1 by Z_2 under this action, G_2 is not stable because the centralizer of σ in G_2 is isomorphic to $GL(2,\mathbb{R})$ and the reals are interpretable in $GL(2,\mathbb{R})$. Thomas had showed this interpretability result and it is contained in the discussion before Lemma 1.9 (if we replace $\mathbb{C}$ by $\mathbb{R}$ and note that the Borel group is definable in GL_2).

It is easy to see [3] that the stability class of a group G with an abelian subgroup A of finite index is determined by the $\mathbb{Z}[G/A]$ (integral group ring) module structure of A and a cocycle map from $G/A \times G/A \to A$. Since G/A is finite the cocycle can be given by naming the finitely many elements in the range. However, Example 3 of [7] shows that the module structure of A may fail to be ω–stable even if A is ω–stable and G/A is finite (indeed $G/A \approx \mathbb{Z}_2 \times \mathbb{Z}_2$). Slight variants will make the module structure not even superstable. Thus to show that every finite extension of $\mathbb{Z}^n$ (for any n) is ω–stable we must rely on some further properties of $\mathbb{Z}^n$.

3.3 Definition. An R module A is *hereditarily κ–stable* if for every ring S such that A admits an S module structure compatible with its R–module structure the S–module A is κ–stable.

The point of the definition is to demand that no S can make an appropriate descending chain of R submodules become definable. This definition leads in the natural way to the notion of structure being hereditarily ω–stable, hereditarily superstable or hereditarily stable. Since [6] showed that the stability class of module is determined by the relevant definable chain conditions (cf. [1] or [17]) the following result is obvious. (S cannot cause subgroups which do not exist to become definable!)

3.4 Lemma. *i) If there is no descending chain of* R*–submodules of* A *then* A *is hereditarily* ω*–stable.*
ii) If there is no descending chain of R*–submodules of* A *such that succesive elements of the chain are of infinite index then* A *is hereditarily superstable.*

Finally we conclude

3.5 Lemma. *For every n, the group* $\mathbb{Z}^n$ *is hereditarily superstable.*

Proof. We will show that $\mathbb{Z}^n$ contains no infinite decreasing chain of subgroups $\langle A_i : i < \omega \rangle$ such that A_{i+1} has infinite index in A_i. Since every subgroup of $\mathbb{Z}^n$ is freely generated by at most n elements (has rank at most n) it suffices to show: If $A \subseteq B \subseteq \mathbb{Z}^n$ and [B:A] is infinite then the rank of A is strictly less than the rank of B. By Lemma 15.4 of [5] we observe the following. For any free abelian group B and subgroup A it is possible to choose bases $b_1,\ldots,b_m$ for B, $a_1,\ldots,a_k$ for A with $k \leq m$ and nonnegative integers n_i such that $a_i = n_i b_i$ and $B/A \approx \oplus\langle b_i \rangle / \langle n_i b_i \rangle$. Thus B/A is infinite just if $k < m$ (i.e. some n_i is zero).

The U–rank of $\mathbb{Z}^n$ is bounded by n since the U–rank can be infinite only if there are arbitrarily long finite chains of subgroups with infinite index at each step. Thus we can deduce

3.6 Conclusion. Every crystallographic group is superstable with finite U–rank.

References

[1] J.T. Baldwin. *Fundamentals of Stability Theory.* Springer–Verlag, 1988.

[2] J.T. Baldwin and A. Pillay. Semisimple Stable and Superstable Groups. to appear in Annals of Pure and Applied Logic.

[3] W. Baur, G.L. Cherlin, and A. Macintyre. Totally categorical groups and rings. *Journal of Algebra,* 57, 407–440, 1979.

[4] G. L. Cherlin. Groups of Small Morley Rank. *Annals of Mathematical Logic,* 17,1–28, 1979.

[5] L. Fuchs. *Infinite Abelian Groups.* Volume 1, Academic Press, New York, 1970.

[6] S. Garavaglia. Decomposition of Totally Transcendental Modules. *The Journal of Symbolic Logic,* 45, 155–164, 1980.

[7] Cherlin G. L. and Rosenstein J. G. On ω_0–Categorical Abelian by Finite Groups. *Journal of Algebra,* 53, 188–226, 1978.

[8] H. Hiller. Crystallography and the Cohomology of Groups. *The American Mathematical Monthly,* 93, 765–778, 1986.

[9] E. Hrushovski. *Contributions to Stable Model Theory.* PhD thesis, University of California, Berkeley, 1986.

[10] E. Hrushovski. Remarks on ω_0-Categorical Structures. submitted.

[11] J. Humphreys. *Linear Algebraic Groups.* Springer–Verlag, 1975.

[12] D. Lascar. Les Groupes ω–Stables de Rang Fini. TAMS 292,1985.

[13] Gary A. Martin. Definability in Reducts of Algebraically Closed Fields. *Journal of Symbolic Logic,* 53, 188–199, 1988.

[14] K. A. Meirembekov. Preservation of Stability Under Finite Extensions of a group. *Algebra and Logic,* 25, 171–182, 1986.

[15] Ali Nesin. Solvable Groups of Finite Morley Rank. Journal of Algebra, to be published.

[16] V. P. Platonov. Proof of the finiteness hypothesis for the solvable sub–groups of Algebraic Groups. *Siberian Journal of Mathematics,* 10, 800–804, 1969. Russian original Sibirskii Mathematicheskii Zhurnal Vol.10 pp. 1084–1090.

[17] Bruno Poizat. *Groupes Stables.* Nur Al–mantiq Wal–ma'rifah, 82, Rue Racine 69100 Villeurbanne France, 1987.

[18] Katsumi Tanaka. Non–Abelian Groups of Morley Rank 2. submitted to Mathematica Japonica.

[19] B. I. Zil'ber. Some Model Theory of Simple Algebraic Groups over Algebraically Closed Fields. *Colloquium Mathematicum,* 48, 173–180, 1984.

NON–ASSOCIATIVE RINGS OF FINITE MORLEY RANK

Ali Nesin*

INTRODUCTION

Stable associative rings have been investigated by Cherlin–Reineke [Ch–Re], by Baldwin–Rose [B–Ro] and by Felgner [Fe] too early in the history of stable algebraic structures to get the attention they deserve. Rose started to investigate stable non–associative rings [Ro] in the late 1970's but again his work did not get the attention of model theorists. Macintyre's classification of ω_1–categorical fields [Mac2] and its generalization to superstable fields [Ch–S] and to ω–stable division rings [Ch2] became important because, we think, of their importance in the study of stable groups. We believe in the near future stable rings (associative or not) and their stable modules will become an important research area in applied model theory. They arise naturally in the study of stable groups. Here we list 4 instances:

a) Zil'ber classified ω_1 – categorical associative rings of characteristic 0 as indecomposable algebras over an algebraically closed field of characteristic 0 [Zi2]. He announces in [Zi3] that the same methods classify also ω_1 – categorical nilpotent Lie algebras over $\mathbb{Q}$ and that using Campbell–Baker–Hausdorff formula (see e.g. [Jac 1]), one can deduce that ω_1 – categorical torsion–free nilpotent groups are algebraic groups over an algebraically closed field of characteristic 0.

b) Let R be an arbitrary ring (not necessarily associative, does not necessarily have a unit). Then on the set $G = R \times R$ we can define a group multiplication by

$$(x,y)\,(x_1,y_1) = (x + x_1,\ y + y_1 + xx_1).$$

This group can also be viewed as

$$G = \left\{ \begin{pmatrix} 1 & x & y \\ 0 & 1 & x \\ 0 & 0 & 1 \end{pmatrix} : x, y \in R \right\}$$

with the obvious multiplication. It is easily checked that G is nilpotent of class ≤ 2 and is Abelian iff R is commutative. Since G is interpretable in R, it inherits

---

* Partially supported by NSF Grant DMS-8801021.

stability properties of R. Thus to classify nilpotent groups, say of small Morley Rank, we should at least classify rings of small Morley Rank.

c) We can generalize the above example. Let R be any ring (not necessarily associative, does not necessarily have a unit) of finite Morley rank. Let M be a left R–module of finite Morley rank. Define

$$G = \left\{ \begin{pmatrix} 1 & r & x \\ 0 & 1 & y \\ 0 & 0 & 1 \end{pmatrix} : r \in R, x, y \in M \right\}$$

with the obvious multiplication. Then G is a nilpotent of class 2 group of finite Morley rank.

If R is associative with an identity and M is an associative, unitary right R–module then $G = M \rtimes R^*$ i.e. the group

$$G = \left\{ \begin{pmatrix} 1 & x \\ 0 & r \end{pmatrix} : r \in R^*, x \in M \right\}$$

has also finite Morley rank. (Here R^* denotes the multiplicative group of invertible elements of R).

d) Associative, commutative rings with identity come naturally into scene when one studies solvable of class 2, centerless, connected groups of finite Morley rank (see [Ne 1]). Such rings of finite Morley rank have been classified by Cherlin and Reineke in [Ch–R]. But it happens that we are also interested in their modules simply because the above groups can be interpretably imbedded into a finite product of groups of the form $M \rtimes R^*$ where M is an R–module of finite Morley rank and R is a ring with the above properties.

The methods of this article are not original. The basic ideas are mainly Zil'ber's. We found analogues of the known results about groups for rings.

In §1 we set the basics to study the non–associative rings. The lemmas are so simple that the mere knowledge of the basic concepts like Morley rank and degree is enough to understand their proofs.

In §2 we discuss Jacobson density theorem for associative rings.

In §3 we show how one can use the density theorem in the study of non–associative rings. The ideas are due to Zil'ber [Zi2]. He applied them to associative rings. We generalize his methods to general rings. In the end of this section we apply the previous results to connected Lie rings of Morley rank 1. Ivo Herzog can prove parts i) and ii) of Lemma 13 by some other methods.

In §4 we study Lie rings of finite Morley rank. We prove the analogue of Zil'ber's theorem for solvable groups. Namely we show that in a connected, solvable, non–nilpotent Lie ring of finite Morley rank one can interpret an algebraically closed field. The method is almost exactly like Zil'ber's original proof. But the result is amazingly different: in the construction of the field K we find K^+ where Zil'ber finds K^*. In other words we find the logarithm of Zil'ber's construction! This may not be so shocking for logicians who know the exp–log correspondance between Lie groups and Lie algebras. We also notice that the construction of this field by means of Zil'ber's methods is also given by Jacobson's density theorem. Finally we notice (without proof) that the proof of [Ne2] can be mimicked to show that if L is solvable and connected then L' is nilpotent.

This article was written while the author was visiting Notre Dame University. He would like to thank the Mathematics department of Notre Dame and especially Julia Knight and Anand Pillay for their hospitality. Thanks are also due to John Baldwin for his mathematical comments and for correcting my English.

§1. BASICS

Before starting to set the basics we would like to recall Zil'ber's indecomposability Theorem [Zi 1] (not in its full generality):

Zil'ber's indecomposability Theorem: Let G be a group of finite Morley rank. Let $(X_i)_{i\in I}$ be a set of connected subgroups of G. Then the group generated by $(X_i)_{i\in I}$ is definable and connected. In fact if H is this group then there are a finite number of $i_1, ..., i_n \in I$ such that

$$H = X_{i_1} \cdot ... \cdot X_{i_n} .$$

If G is Abelian and written additively then we write

$$H = X_{i_1} + \ldots + X_{i_h}.$$

Pillay's article "Model Theory, Stability Theory and Stable Groups" in this volume has a proof of the above Theorem. The reader can also find a proof in [Ne 3] or in [Th].

Unless otherwise stated R will always denote an arbitrary ring of finite Morley rank (not necessarily associative, does not necessarily have an identity). In particular R is a group under addition. Thus it has a connected component R° which is an additive subgroup of R. If I is a definable ideal of R we may also speak about I°, the connected component (as an additive subgroup) of I.

Lemma 1. If I is a left (or right or bi) ideal of R then so is I°.

Proof: For $x \in R$, xI° is a subgroup of R. Since I is a left ideal $xI^\circ \subseteq I$. By Zil'ber's indecomposability theorem the group generated by xI° and I° is connected, thus it is I°, so $xI^\circ \subseteq I^\circ$. □

It follows from Lemma 1 that R° is an ideal of R. We may therefore speak about a "connected" ring. From now on, unless otherwise stated, all rings will be connected. Notice that if $(R,+)$ is torsion–free or divisible then R is necessarily connected.

Lemma 2. If I is a finite left (resp. right) ideal then $RI = 0$ (resp. $IR = 0$).

Proof: Let $i \in I$. R being connected, Ri is a connected additive subgroup of R. But $Ri \subseteq I$, so Ri is finite. A connected finite group is 0. Thus $Ri = 0$. □

For $n \in \mathbb{N}$, define

$$R_n = \{x \in R \mid nx = 0\}.$$

R_n is a definable bi–ideal of R. If n and m are prime to each other, then $R_n \cap R_m = \{0\}$, so also $R_n R_m = 0$. If n divides m then $R_n \subseteq R_m$. For p, a prime number, define

$$R_{p^\infty} = \bigcup_{k\geq 1} R_{p^k}.$$

R_{p^∞} is a bi–ideal. But it is not necessarily definable. Corollary 4 will show that it is almost always definable.

We define $ann_R R = \{r \in R: Rr = 0\}$, $ann R_R = \{r \in R: rR = 0\}$. If R is not associative these are not necessarily ideals of R. But they are definable additive subgroups.

Lemma 3. There is an $n \in \mathbb{N}$ such that

$$p^n R_{p^\infty} \subseteq ann\, R_R \cap ann_R R.$$

Proof: Consider the additive group I generated by Rx, $x \in R_{p^\infty}$. By Zil'ber's indecomposability theorem

$$I = Rx_1 + ... + Rx_k$$

for some $x_1,..., x_k \in R_{p^\infty}$. Let n be such that $x_1,...,x_k \in R_{p^n}$. Then $I \subseteq R_{p^n}$.

In particular for any $x \in R_{p^\infty}$, $Rx \subseteq R_{p^n}$, i.e. $p^n Rx = 0$ or $R p^n x = 0$, i.e. $p^n x \in ann\, R_R$. Similarly $p^n x \in ann_R R$. □

Corollary 4. If $ann\, R_R = 0$ (or $ann_R R = 0$) then R_{p^∞} is definable. In fact $R_{p^\infty} = R_{p^n}$ for some $n \in \mathbb{N}$.

Proof: By Lemma 3, if $x \in R_{p^\infty}$ then $p^n x = 0$, so $x \in R_{p^n}$. □

The next corollary states that the torsion part and the torsion–free part of a ring without annihilator direct sum definably.

Corollary 5. Let R be a connected ring with $ann\, R_R = 0$ then there are finitely many distinct primes $p_1,..., p_k$, there is a definable torsion–free (as an additive group) ideal D such that

$$R = R_{p_1^\infty} \oplus \dots \oplus R_{p_k^\infty} \oplus D.$$

Proof: Notice that by Corollary 4 each $R_{p_i^\infty}$ is definable and has bounded order.

By Macintyre [Mac 1] $R = D \oplus H$ as an additive group where D is divisible and H is of bounded order. If n is such that $nH = 0$ then $D = nR$. So D is definable. This also shows that D is a bi–ideal.

Let $r \in R$ have finite additive order. If $r = d + h$ for $d \in D$, $h \in H$ then $d = r - h \in D$ and has finite order if $d \neq 0$. So if $d \neq 0$, D would have an element d' of prime order p. But then $R_{p^\infty} \neq R_{p^n}$ for any n (because D is divisible). This contradicts Corollary 4. Thus $d = 0$, so $r \in H$. We showed that any element of finite order is in H. Thus $R_{p^\infty} \subseteq H$ for all p.

Clearly any element of finite order can be written as the sum of its primary parts. Thus

$$H = \bigoplus_{p \text{ prime}} R_{p^\infty}.$$

Since R is connected, no R_{p^∞} is finite. Since R has finite Morley rank we can have only finitely many R_{p^∞} involved in H. This proves the Corollary. □

Corollary 5 shows that the study of connected rings with $\text{ann}\, R_R = 0$ can be reduced to the study of their primary parts. Let us underline the essence of Corollary 5:

Proposition 6. Let R be a connected ring with $\text{ann}_R R=0$. Then $R = D \oplus H$ where D, H are definable ideals, D is torsion–free divisible, H has bounded order. □

Lemma 7. Let Σ be a set of connected definable additive subgroups S_i of R. Then the subring and the left (or right, or bi) ideal generated by Σ are definable and connected.

Proof: Define inductively

R_1 = Additive group generated by S_i's,

R_{n+1}= Additive group generated by R_n and xR_n, R_nx for $x \in R_n$.

By induction, using Zil'ber's indecomposability theorem, we see that R_n's are definable and connected. By definition $R_n \subseteq R_{n+1}$. Thus $\bigcup_{n\geq 1} R_n = R_n$ for some n. R_n is clearly the subring generated by Σ. (R_n may not have all the constants of R, e.g. R_n may be without identity even if R has an identity).

To prove the lemma for left ideals, in the definition of R_{n+1} we omit R_nx's and let x range over R. □

It follows from Lemma 7 that if a connected ring R has an identity then every ideal is definable and connected. This is simply because an ideal I is generated by the connected subgroups Rx for $x \in R$. In particular such a ring is Noetherian and Artinian. Rose [Ro] proved that if R is an arbitrary stable ring then J(R), the Jacobson radical of R, is definable and R/J(R) is Artinian and Noetherian.

Corollary 8. R^n, $R^{(n)}$ are definable and connected.

Proof: By definition R^{n+1} is the ideal generated by $\{xy$: for $x \in R, y \in R^n\}$, i.e. the ideal generated by $(xR^n)_{x\in R}$.

Since for x fixed xR^n is a definable homomorphic image of R^n, by induction xR^n is a connected subgroup. Now apply Lemma 7.

For $R^{(n+1)}$ we consider the ideal generated by $(xR^{(n)})_{x\in R^{(n)}}$. The proof is the same. □

Corollary 9. If I is a minimal left ideal of R and if $I \not\subseteq \mathrm{ann}\, R_R$ then I is definable and connected (also infinite by Lemma 2).

Proof: Let $a \in I \setminus \mathrm{ann}\, R_R$. Then $0 \neq Ra \subseteq I$. Thus the left ideal generated by Ra is I which is definable and connected by Lemma 7. □

Remark: If $\omega = \Sigma\, r_n (r_{n-1} (\ldots(r_2r_1) \ldots)$ a formal sum of formal monomials ($r_i \in R$), let us denote by ωa (for $a \in R$) the element of R defined by

$$\Sigma\, r_n(\ldots(r_2(r_1a))\ldots).$$

Then Corollary 9 and the proof of Lemma 7 tell us that the minimal left ideal I is the set of ωa's (a fixed in $I-\{0\}$) for ω ranging over words whose "sum length" and "product length" are bounded by some natural number n. This will be made more precise in §3, Lemma 12.

Corollary 10. The ideal generated by $\{xy - yx: x,y\in R\}$ is definable and connected.

Proof: Let $[x,R] = \{xy - yx \mid y \in R\}$. $[x,R]$ is a definable homomorphic image of R (as an additive group). Thus it is definable and connected. Now the ideal in question is generated by $[x,R]$ for x ranging over R. Use Lemma 7. □

Remark: If C is the ideal generated $\{xy - yx: x,y \in R\}$ then R/C is a commutative ring of finite Morley Rank.

Corollary 11. If $\mathrm{ann}R_R = 0$ then R has minimal left ideals which are definable. Furthermore these minimal ideals I are generated by the set Ra for any fixed $a\in I-\{0\}$, i.e. they are principal ideals.

Proof: By Lemma 7 the left ideal generated by Ra is definable and connected. Choose a minimal such. By Corollary 9 and its proof it is a minimal ideal. □

§2. DENSITY THEOREM FOR ABELIAN GROUPS

In this section we will forget about the multiplicative structure of our ring. We will not assume any stability conditions either. Let R be an Abelian group written additively. An additive group M is said to be an R–module if there is a homomorphism ρ of Abelian groups:

$$\rho: R \to \mathrm{End}_{\mathbb{Z}}(M).$$

M is said to be a faithful module if ρ is injective. We can define a multiplication $rx \in M$ for $r\in R$, $x\in M$ via ρ:

$$rx = \rho(r)(x).$$

Then all the module–theoretical concepts can be defined.

Schur's Lemma: If M is an irreducible R–module then $\Delta = \{\varphi: M \to M:$ φ linear and $\varphi(rx) = r\varphi(x)$ for all $r \in R, x \in M\} = \mathrm{End}_R M$ is a division ring.

Under the conditions of Schur's Lemma, M becomes a vector space over Δ and $\rho(r)$ is a Δ– linear map for $r \in R$. If M is also faithful then R imbeds naturally into $\mathrm{End}_\Delta M$.

A subset S of $\mathrm{End}_\Delta M$ is said to be dense if for any n, any $x_1,\ldots,x_n \in M$ linearly independent over Δ and any $y_1,\ldots,y_n \in M$ there is an $s \in S$ such that

$$s(x_i) = y_i \quad (i = 1, \ldots, n).$$

Notice that if $\dim_\Delta M < \infty$ then a dense subset of $\mathrm{End}_\Delta M$ is necessarily $\mathrm{End}_\Delta M$.

Density theorem for primitive abelian groups: Let M be an irreducible faithful module for the Abelian group R. Define $\Delta = \mathrm{End}_R M$. Then $R \le \mathrm{End}_\Delta M$ as an additive group and the ring S generated by R in $\mathrm{End}_\Delta M$ is dense in $\mathrm{End}_\Delta M$.

Proof: This is a rephrasing of Jacobson's density theorem (see [Jac 2] p. 28) that states the above conclusion in case $R = S$ is an associative ring.

Let S be the (associative) ring generated by R in $\mathrm{End}_\Delta M$. Since $S \subseteq \mathrm{End}_\Delta M$, M is a faithful S–module. Since $R \subseteq S$, M is also S–irreducible. So if $\Delta' = \mathrm{End}_S M$, S is dense in $\mathrm{End}_{\Delta'} M$ by the original Jacobson density theorem. But since $R \subseteq S$ we also have $\mathrm{End}_S M \subseteq \mathrm{End}_R M$. Since S is generated by R, $\mathrm{End}_R M \subseteq \mathrm{End}_S M$. Thus $\mathrm{End}_R M = \mathrm{End}_S M$, i.e. $\Delta = \Delta'$. So S is dense in $\mathrm{End}_{\Delta'} M = \mathrm{End}_\Delta M$. □

If R is a ring then for M to be an R–module we may need to add some more conditions on the R–action. For instance if R has an identity 1 then we want $\rho(1) = \mathrm{Id}_M$. Or if R is an associative ring we want ρ to be a ring homomorphism. If R is a Lie ring so that it satisfies the Jacobi identity $((rs)t + (st)r + (tr)s = 0)$ then we impose to ρ the condition to be a Lie–homomorphism, i.e. $\rho(rs) = \rho(r)\,\rho(s) - \rho(s)\,\rho(r)$.

§3. APPLICATIONS OF THE DENSITY THEOREM

Suppose R is an Abelian group of finite Morley rank. Suppose we can interpret in R a faithful irreducible R–module M (M could be an $\bar{R}$–module where $\bar{R}$ is a group interpreted in R). Suppose also that the division ring $\Delta = End_R M$ is interpretable in R. Then Δ and M have finite Morley rank. It follows that if Δ is infinite then it is an algebraically closed field [Ch 2] and M is a finite dimensional vector space over Δ. Therefore by the density theorem the ring S generated by R in $End_\Delta M$ is $End_\Delta M$! In fact, as John Baldwin noticed, if the ring S generated by R in $End_\Delta M$ is interpretable in R (e.g. if R is already an associative ring) then we do not even need the non–finiteness of Δ to claim that $S = End_\Delta M$. Because in this case M will have finite dimension over Δ anyway: if $x_1,...,x_n,...$ is a Δ–base of M, let

$$S_k = \{s \in S \mid sx_1 = \ldots = sx_k = 0\}.$$

Clearly $S_k \geq S_{k+1}$. But also by the density theorem $S_k \neq S_{k+1}$. This contradicts the descending chain condition. (For this argument we only need stability, because the subgroups S_k are intersections of uniformly definable subgroups).

Therefore to use the density theorem we need the following steps:

1) Find an interpretable faithful irreducible R–module M (or may be $\bar{R}$ –module).

2) Interpret $\Delta = End_R M$ in R.

3) Then we know that $R < End_\Delta M$ (as an additive subgroup) and the ring S generated by R in $End_\Delta M$ is dense in $End_\Delta M$. To show that $S = End_\Delta M$, prove that either Δ is infinite or S is interpretable in R.

For the first step: an obvious candidate for M, in case R is a ring, is a minimal left ideal. Then this ideal will be definable if ann $R_R = 0$ (Corollary 9). The minimality of M will ensure that it is an irreducible module. But we do not necessarily have the faithfulness. Then divide R by the annihilator of M:

$$ann_R M = \{r \in R \mid rM = 0\}.$$

Now $R/ann_R M$ is an additive group and M is an irreducible and faithful $R/ann_R M$–module. Notice that $ann_R M$ is not necessarily an ideal. But it is so if R is an associative or a Lie ring (see end of the section for the definition of Lie ring).

Now about the second step; it is astonishing that Δ is almost always interpretable in R.

Lemma 12. Let M be a minimal left ideal (necessarily definable) in a connected ring of finite Morley rank with $\mathrm{ann}R_R = 0$. Then $\Delta = \mathrm{End}_R M$ is an interpretable division ring (hence a field). If Char R = 0 or more generally if Δ is infinite then Δ is an algebraically closed field.

Proof: (The idea of the proof is from [Zi2]). We know by Corollary 9 that M is definable. We need to recall explicitly its definition. Let ω be a formal word in R of the form

$$\omega = \omega_1 + \ldots + \omega_n,$$
$$\omega_j = r_{j1}\,(r_{j2}(\ldots(r_{jk}\,r_{j,k+1})\ldots)).$$

Here, k depends on j. Such words will be called special. For $a \in R$ and ω a special word, define an element $\omega(a)$ of R by

$$\omega(a) = \omega_1(a) + \ldots + \omega_n(a),$$
$$\omega_j(a) = r_{j1}\,(r_{j2}(\ldots(r_{jk}(\,r_{j,k+1}a))\ldots)).$$

Notice that $\omega(a) \in R$ and is not a formal word. Let us also define the length $\ell(\omega)$ of a special word ω by

$$\ell(\omega) = \ell(\omega_1) + \ldots + \ell(\omega_n),$$
$$\ell(\omega_j) = k + 1 \text{ (see the definition of } \omega_j).$$

Now we are ready to give the explicit definition of M. Let $a_o \in M-\{0\}$. Then for some integer n,

$$M = \{\omega(a_o) : \omega \text{ is a special word of length } \leq n\}.$$

This is the content of the proof of Lemma 7. Fix such an integer n.

We need one more definition before interpreting Δ. Let

$$\begin{aligned} J = \{a \in M:\ & \forall\, \omega, \omega_1, \omega_2 \text{ special words of length } \leq n,\\ & (\omega(a_o) = \omega_1(a_o) \rightarrow \omega(a) = \omega_1(a))\\ & \& \ (\forall r \in R \ \ r(\omega_1(a_o)) = \omega(a_o)) \rightarrow r(\omega_1(a)) = \omega(a))\\ & \& \ (\omega_1(a_o) + \omega_2(a_o) = \omega(a_o) \rightarrow \omega_1(a) + \omega_2(a) = \omega(a))\}. \end{aligned}$$

J is a definable subset of R and $a_o \in J$. Clearly J is an additive subgroup of R. Thus J will be infinite if charR = 0.

Now for $a \in J$ define $\gamma_a: M \to M$ by

$$\gamma_a(\omega(a_o)) = \omega(a).$$

The definition of J implies that γ_a is well defined and is in $\mathrm{End}_R M$. Conversely if $\sigma \in \mathrm{End}_R M$ then clearly $\sigma(a_o) \in J$ and $\sigma = \gamma_{\sigma(a_o)}$. Thus

$$\Delta = \mathrm{End}_R M = \{\gamma_a: a \in J\}$$

is interpretable in R. □

Now we can carry out our third step:

Corollary 13. Let R be a connected ring of finite Morley rank with $\mathrm{ann}\, R_R = 0$. Let M be a minimal left ideal of R (Corollary 11). Then

i) M is necessarily definable (Corollary 9).

ii) M is a faithful irreducible $R/\mathrm{ann}_R M$–module. Let $\bar{R} = R/\mathrm{ann}_R M$.

iii) $\Delta = \mathrm{End}_{\bar{R}}(M) = \mathrm{End}_R M$ is interpretable in R and is a field (algebraically closed if infinite).

iv) $R < \mathrm{End}_\Delta M$ as an additive group and the ring S generated by R in $\mathrm{End}_\Delta M$ is dense in $\mathrm{End}_\Delta M$.

v) If either char $R = 0$ or S is interpretable in R or S is commutative then Δ is infinite and so M is a finite dimensional vector space over Δ. Hence $S = \mathrm{End}_\Delta M$. Also if S is commutative then $\dim_\Delta M = 1$.

Proof: Everything is already proved except some parts of v). If S is interpretable we noticed in the beginning of this section that M must have finite dimension over Δ. So if Δ is finite then M is also finite. But then by Lemma 2 $M \subseteq \mathrm{Ann} R_R = 0$, a contradiction.

If S is commutative then (since it is dense in $\mathrm{End}_\Delta M$) it can easily be checked that $\dim_\Delta M = 1$. Again Δ is infinite. □

Conjecture: $\dim_\Delta M < \infty$ always (notation as above).

Let us give an illustration of this Corollary.

Recall that a Lie ring is an additive group L with a bilinear product (called bracket) $[x,y]$ such that for all $x,y,z \in L$

$[x,x] = 0$,

$[[x,y], z] + [[y,z], x] + [[z,x], y] = 0$ (Jacobi identity).

Sometimes we will omit the brackets and write xy for $[x,y]$.

Let L be a connected Lie ring of Morley rank 1. We would like to prove that L is Abelian, i.e.

$$[x,y] = 0$$

for all $x,y \in L$. But being unable to prove it, let us see what the Corollary gives us.

Define $Z(L) = \{x \in L : [x,y] = 0\}$ = center of L,

$Z_2(L) = \{x \in L : [x,y] \in Z(L)\}$.

By the Jacobi identity $Z(L)$, $Z_2(L)$ are ideals. If $Z(L)$ is infinite then $Z(L) = L$ and so L is abelian. Suppose therefore that $Z(L)$ is finite. Let $\bar{L} = L/Z(L)$. Then $Z(\bar{L}) = Z_2(L)/Z(L)$. Suppose $Z_2(L)$ is infinite. Then $Z_2(L) = L$. Thus $L^2 \subseteq Z(L)$. But by Corollary 8, L^2 is connected. Thus $L^2 = 0$, $Z(L) = L$, a contradiction. Thus $Z_2(L)$ is finite. Then by Lemma 2, $Z_2(L) \subseteq Z(L)$. Thus $Z_2(L) = Z(L)$ and $\bar{L}$ is centerless.

We showed the following:

Lemma 14: If L is a connected non Abelian Lie ring of Morley rank 1 then $\bar{L} = L/Z(L)$ is a connected centerless Lie ring of Morley rank 1. □

Now we can apply the Corollary to $\bar{L}$. Assume $L = \bar{L}$ for the sake of notational simplicity. ann $L_L = 0$ (because $Z(L) = 0$), $M = L$ (because L has Morley rank 1, so is a minimal (definable) ideal). So we have parts i) and ii) of the following Lemma.

Lemma 15. If L is a centerless connected Lie ring of Morley rank 1 then

i) $L \subseteq \mathrm{End}_\Delta L$ where $\Delta = \mathrm{End}_L L$.

ii) The associative ring generated by L in $\mathrm{End}_\Delta L$ is dense in $\mathrm{End}_\Delta L$.

iii) L has no non–trivial ideals (i.e. L is simple).

iv) Δ is a finite field.

v) L has characteristic p for some prime $p \neq 0$.

Proof: We have already proved i) and ii). If I were a proper ideal of L then the ideal J generated by $\{[L,x]: x \in I\}$ would be definable by Lemma 7. Since this definable ideal J is in $I \subsetneq L$, J would be finite and hence central by Lemma 2. Thus $J = 0$. Then $Lx = 0$ all $x \in I$, i.e. $I \subseteq Z(L) = 0$. This proves iii).

Let $a \in L \setminus \{0\}$. Let $C_L(a) = \{x \in L \mid [x,a] = 0\}$. Since $\Delta = End_L L$, Δ acts on the finite (but non–zero) set $C_L(a)$. If Δ were infinite it would be an algebraically closed field and so it would be connected, then $\Delta a \subseteq C_L(a)$ would also be connected. But $C_L(a)$ is finite, so $\Delta a = 0$. Since $Id \in \Delta$, this is a contradiction. This proves iv).

If v) were not true then $nx \in C_L(x)$ for all $n \in \mathbb{N}$, so $C_L(x) = L$, $x = 0$. □

Let us make a weaker conjecture then the previous one:

Conjecture: Connected Lie rings of Morley rank 1 are Abelian.

Reineke proved (see [Re] or [Ch 1]) that connected groups of Morley rank 1 are Abelian. The proof is very easy but one cannot give the same proof for Lie rings. Cherlin and the author proved that if L is non–Abelian then $\dim_\Delta C_L(a) > 1$ for a generic element a of L.

Added to the last version: In view of Hrushovski's discovery of new strongly minimal sets the author of the above conjecture does not believe in it anymore, thus:

Conjecture: There is a non–Abelian connected Lie ring of Morley rank 1.

§ 4. SOLVABLE, NON–NILPOTENT LIE RINGS

Let us first recall the definitions of solvable and nilpotent rings. We defined the ideals R^n and $R^{(n)}$ in Corollary 8. A ring R is said to be solvable if $R^{(n)} = 0$ for some n. It is said to be nilpotent if $R^n = 0$ some n. Nilpotent implies solvable.

Let us also define the centers: $Z_o(R) = 0$,

$Z_{i+1}(R) = \{x \in R : xR \subseteq Z_i(R)\}$.

If $R = L$ is a Lie ring then by the Jacobi identity $Z_i(L)$ is an ideal of L. Clearly $Z_i(L) \subseteq Z_{i+1}(L)$. It is relatively easy to check that L is nilpotent iff $Z_m(L) = L$ for some m. We have $Z_1(L) = Z(L) = ann_L L$. Notice also that $Z(L/Z_i(L)) = Z_{i+1}(L)/Z_i(L)$.

We should also remind the reader that in a Lie ring we have $[x,x] = 0$. This implies $[x,y] = -[y,x]$. So that all left (or right) ideals are bi–ideals.

If $X, Y \subseteq L$ are any subsets, then the centralizer of X in Y is $C_Y(X) = \{y \in Y \mid [x,y] = 0 \text{ for all } x \in X\}$.

Theorem 16. Let L be a connected, solvable, non–nilpotent Lie ring of finite Morley rank. Then an algebraically closed field can be interpreted in L.

Proof: We will follow Zil'ber's steps (see [Zi 1] or [Ne 3] or [Th]). Since L is connected and not nilpotent it is infinite.

We first reduce the problem to the case where L is centerless:

Claim 1: Without loss of generality L is centerless.

Divide L by its centers until there isn't any left. The point is that we need to divide L only a finite number of times. Since L has finite Morley rank and is not nilpotent and since to divide by an infinite definable ideal decreases the Morley rank, at some point we can only divide by finite ideals Z_{i+1}/Z_i. Assume without loss of generality that $Z = Z(L)$ and $Z_2 = Z_2(L)$ are finite. Then by Lemma 2, $Z_2 \subseteq Z$. Thus $Z_2 = Z$. □

From now on we assume that L is centerless. By Lemma 2 this implies that L has no, non–zero, finite ideals.

Let A be a minimal ideal of L. A exists and is definable by Corollary 11 ($\text{ann}\, R_R = Z(L) = 0$). By Lemma 1, A is connected. By Corollary 8, A^2 is definable. Since L is solvable, so is A. Thus $A^2 \subsetneq A$. But then A^2 is finite, so $A^2 = 0$. This shows that A is Abelian, i.e. $[A,A] = 0$.

Let $C = C_L(A) = \{x \in L: [x,A] = 0\}$. Since A is an ideal of L, so is C. L being centerless, $C \subsetneq L$. L being connected L/C is an infinite Lie ring. Let H be a minimal definable ideal of L such that $C \subsetneq H \subseteq L$ and H/C is infinite. Since $[L,A] \subseteq A$, also $[H,A] \subseteq A$. Choose $B \subseteq A$, a minimal (definable) infinite ideal such that $[H,B] \subseteq B$. B is again connected.

Claim 2: $C_A(H) = 0$, $C_B(H) = 0$.

Since A and H are ideals, $C_A(H)$ is an ideal of L. By minimality of A, $C_A(H)$ is either A or finite. If it is A then $H \subseteq C_L(A)$, a contradiction. So it is finite. Then it is 0 because finite ideals of L are 0. Since $C_B(H) \subseteq C_A(H)$, the second equality follows from the first one. □

Claim 3: $H/C_L(A)$ is connected.

Let $H°$ be the connected component of H. $H°$ is still an ideal by Lemma 1. We have $C_L(A) \subseteq H° + C_L(A) \subseteq H$. Since $H/H°$ is finite so is $H/H° + C_L(A)$. Thus $H° + C_L(A)/C_L(A)$ is infinite. Hence by the choice of H, $H = H° + C_L(A)$. But now

$$H/C_L(A) = H° + C_L(A)/C_L(A) \approx H°/H° \cap C_L(A).$$

Since $H°$ is connected, so is $H°/H° \cap C_L(A)$ and therefore also $H/C_L(A)$. □

Claim 4: $H/C_L(A)$ is abelian, i.e. $[H,H] \subseteq C_L(A)$.

By Claim 3 and Corollary 8 $(H/C_L(A)^2$ is connected. But it is also finite (because $H/C_L(A)$ is solvable as L is and it has no infinite definable ideals). Thus $(H/C_L(A))^2 = 0$, i.e. $H^2 \subseteq C_L(A)$. With the notation of Lie rings: $[H,H] \subseteq C_L(A)$. □

Claim 5: If $h \in H$ then $C_B(h) = 0$ or B.

Claim 4 and Jacobi identity imply that $C_B(h)$ is an H–ideal. If $C_B(h)$ is not finite, then it is B by the minimality of B. If it is finite, since the connected ring $H/C_L(A)$ acts on it, as in Lemma 2, $H/C_L(A)$ annihilates $C_B(h)$. So also H annihilates $C_B(h)$. Thus $C_B(h) \subseteq C_B(H) = 0$. □

Claim 6: $H/C_H(B)$ is infinite.

$C_H(B) = C_L(B) \cap H \supseteq C_L(A) \cap H = C_L(A)$. Thus $H/C_H(B) = (H/C_L(A))/(C_H(B)/C_L(A))$. So $H/C_H(B)$ is connected by Claim 3. Now if $H/C_H(B)$ were finite then we would have $C_H(B) = H$, or $B \subseteq C_A(H)$, contradicting Claim 2.□

By Claim 6 the infinite Lie ring $H/C_H(B)$ acts on B by adjoint representation:

$$\begin{array}{ccc} H/C_H(B) & \longrightarrow & \mathrm{End}\, B \\ \bar{h} & \longrightarrow & \mathrm{ad}\, h \end{array}$$

where $(\text{ad } h)(b) = [h, b]$ for $h \in H$, $b \in B$.

By Claim 5 adh is a 1–1 map if $h \neq 0$. Since B has finite Morley rank it is also onto. Thus we have an imbedding

$$H/C_H(B) \to \text{Aut } B \cup \{0\}$$

Let R be the (associative) subring generated by the image of $H/C_H(B)$ in End B. Since $[H,H] \subseteq C_H(A) \subseteq C_H(B)$, R is a commutative ring. We will show that R is an algebraically closed field. By Claim 6, R is infinite.

Let $b \in B - \{0\}$ be a fixed element. As in the proof of Lemma 11 there is a natural number n for which

$$B = \{\omega(b) \mid \omega \text{ is a special word with entries in } H \text{ of length } \leq n\}.$$

This is because B is minimal H–normal so the H–ideal generated by $[H,b]$ must be B.

This says that if $R_1 \subseteq R$ is the set of endomorphisms of "length $\leq n$" then $R_1(b) = B$.

Claim 7: Let $r \in R$, $c \in B - \{0\}$. If $r(c) = 0$ then $r = 0$.

Without loss of generality $b = c$. Then $0 = R_1(r(b)) = r(R_1(b)) = r(B)$. So $r = 0$. □

Claim 8: $R = R_1$, i.e. R is interpretable.

Let $r \in R$. Then $\exists r_1 \in R_1$ such that $r(b) = r_1(b)$. By claim 7, $r = r_1$. □

Now we are ready to show that R is a field. By Claim 7, R has no zero–divisors. By Claim 7 again, there is a 1–1 correspondance between the elements of B and the elements of R:

$$R \to B$$
$$r \to r(b).$$

This is a homomorphism which is 1–1 and onto. Thus the additive group of R and B are isomorphic. In particular R is connected. So if $r \in R - \{0\}$, $Rr = R = rR$. Therefore there is a $u \in R$ such that $ur = r$. Now if $s \in R$ then for some $t \in R$, $s = rt$. So

$$us = u(rt) = (ur)t = rt = s.$$

Thus u is an identity of R.

Let us show that inverses exist. If $r \in R-\{0\}$ and $b' = r(b)$, applying to b' what we have said about b ($b' \neq 0$) we get an $s \in R$ such that $s(b') = b$. So $sr(b) = b = u(b)$. Therefore $sr = u$.

Thus R is an infinite field which is interpretable in L. By Macintyre [Mac 2] it is algebraically closed. Theorem 16 is now proved. □

Let us have a closer look at our field R. We have seen that $(R,+) \simeq (B,+)$ as additive groups. This is also the case in the construction of a field in a solvable connected group. Also in the above case $\mathrm{ad}\, H \subseteq \mathrm{Aut}\,(B) \cup \{0\}$ is an additive subgroup of R and it generates R as a ring. On the other hand in the case of solvable connected groups $\mathrm{ad}\, H$ (H acting by conjugation) is a multiplicative group!

There is a conjecture (which is part of another conjecture called Zil'ber's conjecture) that states that an infinite field R of finite Morley rank cannot have a proper infinite definable subgroup (additive or multiplicative). If this conjecture is true for additive subgroups then $\mathrm{ad}\, H = R$ and $B = [H,b]$ for any $b \in B \setminus \{0\}$! Since the conjecture for additive subgroups is true in case $\mathrm{Char} R = 0$ we have

Corollary 17. (Notation as in the proof of Theorem 16) If $\mathrm{Char}\, L = 0$ then $H/C_H(B)$ and B are isomorphic as additive subgroups via

$$h \to [h,b]$$

for any fixed element $b \in B - \{0\}$. If we define a multiplication on $H/C_H(B)$ by

$$h_1 * h_2 = h \iff [h_1\, [h_2,b]] = [h,b]$$

then $(H/C_H(B), +, *)$ is an algebraically closed field of characteristic 0. □

We know that if G is a connected solvable group of finite Morley rank then $G' = [G,G]$ is nilpotent ([Zi 4], [Ne 2]). We presume that one can mimic the proof in [Ne 2] to show the following:

Theorem 18. If L is a connected solvable Lie ring of finite Morley rank then $L^2 = [L,L]$ is nilpotent.

Now we will use the density theorem once more.

Theorem 19. Let L be a connected solvable centerless Lie Ring of finite Morley rank. Let $A \subseteq Z(L^2)$ be a minimal ideal (then A is infinite and is definable). Let

$\Delta = End_L(A)$. Then $dim_\Delta A = 1$ and the ring generated by $L/C_L(A)$ in $End_\Delta(A)$ is isomorphic to Δ.

Proof: We know by Corollary 13 that the ring generated by $L/C_L(A)$ in $End_\Delta A$ is dense in $End_\Delta A$. But since $A \subseteq Z(L^2)$ we have, $L^2 \subseteq C_L(A)$. So $L/C_L(A)$, hence S, are commutative. But a commutative ring can be dense in $End_\Delta A$ iff $dim_\Delta A = 1$. Thus $S = End_\Delta A \simeq \Delta$. □

The field S we get in this way is the field we got in Theorem 16. To convince yourself of this fact, trace back the definition of $\Delta = End_L(A)$ and notice that it is just the ring R of Theorem 16.

Bibliography

[B–Ro] J. Baldwin and B. Rose, "ω_0 – Categoricity and Stability of Rings" J. Algebra 45 (1977), 1–16.

[Ch 1] G. Cherlin, "Groups of small Morley Rank", Annals Math. logic, 17 (1979), 1–28.

[Ch 2] G. Cherlin, "Superstable Division Rings", Logic Colloquim '77. Proceedings ASL European Summer Meeting, Wroclaw, Poland. Ed. by A. Macintyre, L. Pacholski, J. Paris, North Holland Publishing Company, 1978.

[Ch–Re] G. Cherlin – J. Reineke, "Stability and Categoricity for Commutative Rings", Annals Math Logic 10 (1976), 376–399.

[Ch–S] G. Cherlin–S. Shelah, "Superstable Fields and Groups" Annals Math Logic 18 (1980), 227–270.

[Fe] U. Felgner, "ω_1–Kategorishe Theorien nicht–Kommutativen Ringe", Fund Math 82 (1975), 331–346.

[Jac 1] N. Jacobson, Lie Algebras , Denver Publications, (1962).

[Jac 2] N. Jacobson, Structure of Rings, AMS Colloquim Publications, vol. XXXVII (1964).

[Mac 1] A. Macintyre, "On ω_1–Categorical Theories of Abelian Groups", Fund. Math. 70 (1971), 253–270.

[Mac 2] A. Macintyre, "On ω_1–Categorical Theories of Fields", Fund. Math. 71 (1971), 1–25.

[Ne 1] A. Nesin, "On Solvable Groups of finite Morley Rank ". Submitted to Trans. of AMS.

[Ne 2] A. Nesin, "Solvable Groups of finite Morley Rank" to be published in J. Algebra (1988 +).

[Ne 3] A. Nesin, "Zil'ber's Results", Preprint.

[Re] J. Reineke, "Minimale Gruppen" S. Math. Logik Grundlagen Math 21 (1975), 357–379.

[Ro] B.I. Rose, "Model Theory of Alternative Rings", Notre Dame Journal of Formal Logic, vol XIX, number 2, April 1978, 215–243.

[Th] S. Thomas, Preprint.

[Zi 1] B. Zil'ber, "Groups and Rings with Categorical Theories" (in Russian) Fund. Math 95 (1977), 173–188.

[Zi 2] B. Zil'ber, "Rings with ω_1–Categorical theories", Algebra i Logika, vol 13, no 2, (1974), 168–187. (Translation exists).

[Zi 3] B. Zil'ber, "Uncountable Categorical nilpotent groups and Lie Algebras" lzvestiya VUZ. Matematika 26, no 5, (1982), 75.

[Zi 4] B. Zil'ber, "Groups with Categorical Theories" Abstract of papers presented at the Fourth All–Union Symposium on Group Theory, Novosibirsk, 1973. In Russian, Math inst. Sibirsk. Otdel. Aikad. Nauk SSSR, pp 63–68.

MODULES WITH REGULAR GENERIC TYPES

Ivo Herzog and Philipp Rothmaler

Part I. The Basics

Part II. The Connected Case

Part III. The Ziegler Spectrum

Parts I and II

Philipp Rothmaler

Посвящается памяти моего
дорогого друга Дато Давлианидзе

For structures having a definable group action the notion of generic type makes sense. Poizat showed that having regular generics is a particularly useful model-theoretic property. In this series of papers we try to provide the reader with a systematic treatment of this in the case of modules, where - naturally - addition is taken as the corresponding group action.

I apologize for the long list of acknowledgements following in order reverse to chronological. But I experienced a lot of help and friendship in the last couple of years, which has to do with this paper in one way or another, and which I don't want to leave unmentioned. First, I would very much like to thank my colleagues and friends at Notre Dame and in Indiana for their hospitality and inspiration during the summer of 1987 without which this article would have never been written. Also I should like to thank them for letting me have a wonderful time when visiting Notre Dame during the Fall Semester of 1985 without which I would never have gone back there last summer. Second, I thank the organizers of the Paris Logic Colloquium '85 for giving me a grant so I could present there part of these results. Further, I wish to thank my British Colleagues together with their families, whose generosity I enjoyed in Spring 1985 (when the notion of finitizer occurred to me).

Last, but not least, I thank my colleagues at the Christian-Albrechts-Universität zu Kiel and the Deutsche Forschungsgemeinschaft for their helpful support.

Part I. Basics.

The purpose of this part is twofold. First it is meant to serve as a uniform base, introducing all that is needed in the sequel. This concerns both, notation and previous results. Second it should acquaint the reader with the class of modules which serves as the title for the entire article.

§0. Preliminaries.

Following Weglorz, the language we work in is L_R whose non-logical symbols are 0, +, and a unary function symbol r for every element r of the (once for all fixed, associative, unitary) ring R with the obvious interpretations. The class of all left R-modules can easily be axiomatized in L_R by a certain theory T_R. Throughout, "module" means "model of T_R". A positive primitive (pp) formula is one of the form $\exists \overline{y} \bigwedge_{i<n} (s_i(\overline{x}) + t_i(\overline{y}) = 0)$, where s_i and t_i are R-terms (or, more exactly, L_R-terms), i.e. linear expressions with coefficients from R. In a module M these define projections of solution sets of finite systems of linear equations over R, which are thus subgroups of $M^{\ell(\overline{x})}$ (here ℓ stands for length). By a pp subgroup I (mostly) mean such a definable set where $\ell(\overline{x}) = 1$. I will often not distinguish between formulas and the sets they define. So I allow myself to talk e.g. of a "pp subgroup $\varphi(x)$". Similarly, I will write $\psi \subseteq \varphi$ instead of $T_R \vdash \psi \rightarrow \varphi$. Writing " φ/ψ" (or "$\varphi/\psi(M)$") I mean a pair φ,ψ of 1-place pp formulas with $\psi \subseteq \varphi$ (or its factor group $\varphi(M)/\psi(M)$). I will say " φ/ψ is finite in M" (or in T) if $\varphi/\psi(M)$ is finite (for each (=some) $M \models T$). It is easily seen that $\varphi/\psi(A \oplus B) \cong \varphi/\psi(A) \oplus \varphi/\psi(B)$ (as abelian groups) and thus $|\varphi/\psi(A \oplus B)| = |\varphi/\psi(A)| \cdot |\varphi/\psi(B)|$.

For cardinals I will adopt the following convention: $\kappa = \lambda \pmod{\infty}$ iff $\kappa = \lambda$ or both κ and λ are infinite. Then a necessary and sufficient condition for given modules M and N to be elementarily equivalent is: $|\varphi/\psi(M)| = |\varphi/\psi(N)| \pmod{\infty}$ for all φ/ψ.

This is due to Baur, Garavaglia, and Monk; see footnote at the end of [Ga] for the history of this result. For more historical background I refer the reader to [Zie] and [PR] (see also [Ro 4]). The general theory of pure injective (p.i.) modules, p.i. indecomposables, p.i. hulls, indecomposable types etc. can also be found in these sources.. For easier reference I will quote a few below.

As $|\varphi/\psi(M)|$ is an invariant of T (which is always assumed to be complete) (mod ∞), I may define $\varphi/\psi(T)$ to be n if $|\varphi/\psi(M)| = n$ for some $M \models T$, and ω otherwise. The aforementioned criterion then shows that T is axiomatized by the collection of all the $\varphi/\psi(T)$'s.

If p is a type, p^+ denotes the collection of instances of pp formulas in p. Baur's quantifier elimination theorem implies that in T two complete types p and q over the same set coincide iff $p^+ = q^+$.

If p is in S(A), p^- denotes the collection of all pp formulas over A whose negation is in p.

Notice, $|T| = |R| + \aleph_0$ and $|T|^+$-saturated modules are p.i.

The following lemma due to B. H. Neumann will be refered to as BHN: If a coset of a group A is contained in finitely many cosets of the groups $B_0,...,B_{n-1}$, then all of those where $A/A \cap B_j$ is infinite can be omitted. In particular, there is some j in n such that $A/A \cap B_j$ is finite.

Most of the notation is standard or borrowed from [Zie]. The p.i. hull of A is denoted by H(A). The complete type of A over B by t(A/B). $t^+(A/B)$ stands for $t(A/B)^+$.

If $\bar{a}_j$ are in N_j, $\bar{a} = \bar{a}_1 + \bar{a}_2$ is in $N_1 \oplus N_2$, then clearly $t^+(\bar{a}) = t^+(\bar{a}_1) \cap t^+(\bar{a}_2)$. From [PR, Ch.4, Sect.4] recall

Fact 1. If $H(\bar{a}) = N_1 \oplus N_2$ and $N_1, N_2 \neq 0$ then for a corresponding decomposition $\bar{a} = \bar{a}_1 + \bar{a}_2$ both $t^+(\bar{a}_1)$ and $t^+(\bar{a}_2)$ strictly contain $t^+(\bar{a})$. This was used in the course of proof of

Fact 2. [Zie, 4.4] p from S(0) is indecomposable (which means that H(p) = $H(\bar{a})$ is indecomposable, where $\bar{a}$ realizes p) iff for all $\psi_1, \psi_2 \in p^-$ there is some $\phi \in p^+$ such that $(\psi_1 \wedge \phi) + (\psi_2 \wedge \phi) \in p^-$.

Following Ziegler I write $\phi/\psi \in p$ instead of "$\phi \in p^+$ and $\psi \in p^-$ ".

Fact 3. [Zie, 7.10] Let ϕ/ψ be in the indecomposable types p and q from S(0). If $H(p) \neq H(q)$ then there is a pp $\chi \subseteq \phi$ containing ψ such that $\phi/\chi \in p$ and $\chi/\psi \in q$ or vice versa.

As common in model theory, when dealing with a complete theory T, I will work in a highly saturated (hence p.i.) universal domain, the so-called monster model of T, which contains every set I am working with. By a model I then mean an elementary substructure of the monster. In particular, any p.i. model is a direct summand of the monster, as is each p.i. submodule. Consult [SH] for this habit.

Set-theoretic inclusion is denoted by $\subseteq$, whereas $\subset$ will denote proper inclusion.

Algebraic notation mainly follows [ST]. In particular, a domain is a not necessarily commutative ring without zero divisors. R^o denotes $R \setminus 0$.

§1. The unlimited part.

The theory of unlimited types was developed by Prest. In this section I will briefly present it in the modified form given in [B-R], from which the first fact is taken. (I will work in this section in a fixed completion T of T_R; the formulas about direct sums stated in the preceding section will be used without mention).

Lemma 1. (1) There is a complete theory T_U, the so-called unlimited part of T, satisfying

$$\phi/\psi(T_U) = \begin{cases} 1, \text{ if } \phi/\psi(T) < \omega \\ \omega, \text{ if } \phi/\psi(T) = \omega \end{cases}$$

for all ϕ/ψ. In particular, $(T_U)_U = T_U$.

(2) A is a direct summand of a model of T_U

iff $M \equiv M \oplus A$ for any (some) model M of T

iff $U \equiv U \oplus A$ for any (some) model U of T_U

iff $\phi/\psi(T) < \omega$ implies $\phi/\psi(A) = 0$.

Thus the class of all unlimited summands, i.e. of all summands of models of T_U, is axiomatizable, too.

Proof: To prove T_U as given is consistent and complete, choose M,N,U satisfying (we called such a triple beautiful in [B-R]):
$M < N \models T$, M is $|T|^+$-saturated (hence p.i.), N realizes all types over $M \cup \bar{a}$ for any $\bar{a}$ from N, $U \simeq N/M$. Clearly we can write $N = M \oplus U$. Then $|\phi/\psi(M)| = |\phi/\psi(N)| = |\phi/\psi(M)| \cdot |\phi/\psi(U)|$, hence $|\phi/\psi(U)| = 1$ if $\phi/\psi(T)$ is finite. If not, $\{ \phi(x) \wedge \neg\psi(x\text{-}a_i\text{-}c)\colon c \varepsilon M, i<n\}$ is consistent for each set of representatives $\{a_i : i<n\}$ of ϕ/ψ in U, which can therefore be extended, by choice of N. Thus $T_U = Th(U)$.
(2) Clearly, $M \equiv M \oplus A$ iff $|\phi/\psi(M)| < \omega$ implies $|\phi/\psi(A)| = 1$. The same is true for $U \models T_U$, whence (1) yields $M \equiv M \oplus A$ iff $U \equiv U \oplus A$. Thus it is enough to work with T_U. By (1), $\phi/\psi(T_U) \varepsilon \{1,\omega\}$, so $A \oplus B \models T_U$ and $\phi/\psi(T_U) < \omega$ imply $\phi/\psi(A \oplus B) = 0$, hence $\phi/\psi(A) = 0$. If, on the other hand, this condition holds, A is a direct summand of $U \oplus A \models T_U$. □

This implies that T_U is <u>closed under products</u>, i.e. $U \oplus U' \models T_U$ whenever U and U′ are models of T_U (or, equivalently, when $\phi/\psi(T_U) \varepsilon \{1,\omega\}$ for all ϕ/ψ).

As in [B-R], a pp type p over 0 is said to be <u>closed under finite index</u> if any ψ is in p^+ if it is a pp subgroup of finite index in some $\varphi \in p^+$.

Call p in S(0) <u>unlimited</u> if p^+ is closed under finite index. <u>$S_*(0)$</u> is the set of all unlimited types from S(0).

Next I will show that the unlimited types are exactly those which are realized in the unlimited part, and further that there is no loss in restricting to T_U when talking about unlimited types of T.

Following Prest set

$$p_*^+ = \{\psi(\bar{x}) : \psi \text{ is pp and there is } \phi(\bar{x},\bar{a}) \varepsilon p^+ \text{ with } \phi(\bar{x},\bar{0})/\psi(\bar{x}) \text{ finite}\}.$$

Clearly $p \varepsilon S(0)$ is unlimited iff $p^+ = p_*^+$.

Types over models have a similar feature:

Lemma 2. If $p \varepsilon S(M)$, where M is a model, then $p_*^+ = \{\phi(\bar{x})\colon \varphi(\bar{x})$ is $\phi(\bar{x},\bar{0})$ for some $\phi(\bar{x},\bar{m}) \varepsilon p^+\}$.

Proof: Let $\phi(\overline{x},\overline{m}) \,\varepsilon\, p^+$ and ϕ/ψ be finite. We have to show, p contains some formula defining a coset of ψ. There is a number n such that $M \models$ $\exists\overline{x}_0...\overline{x}_{n-1}\forall\overline{x}\ (\bigwedge_{i<j<n} [\phi(\overline{x}_i,\overline{m}) \wedge \neg\psi(\overline{x}_i-\overline{x}_j)] \wedge [\phi(\overline{x},\overline{m}) \rightarrow \bigvee_{i<n} \psi(\overline{x}-\overline{x}_i)])$.

This means, there are representatives in M for each coset of ψ which is contained in $\phi(M,\overline{m})$. Thus p must contain one of these. □

That every p_*^+ comes from some unlimited p_* (thus justifying the notation) will be established after the next lemma, which is also taken from [B-R].

Lemma 3. (1) Every filter of pp subgroups which is closed under finite index (in the above sense) is the pp part of a uniquely determined (unlimited) type from S(0).

(2) Every filter of pp subgroups in a theory closed under products is the pp part of a uniquely determined type from S(0).

Proof: (2) follows from (1). Let r be a filter as in (1). The only thing to prove is that $r \cup \{\neg\psi : \psi \text{ is pp and } \psi \notin r\}$ is consistent. As r is closed under finite conjunction, it suffices to verify that $\phi \wedge \bigwedge_{i<n} \neg\psi_i$ is consistent with T for any $\phi \,\varepsilon\, r$ and pp $\psi_i \notin r$. If not, $\phi \subseteq \bigvee_{i<n} \psi_i$. Then, by BHN, some $\phi/\phi\wedge\psi_i$ must be finite, whence $\phi\wedge\psi_i \,\varepsilon\, r$, contradicting the choice of ψ_i. □

Corollary 4. [Pr] For all $p \in S(A)$ there is a unique type $p_* \in S(0)$, the so-called <u>free part</u> of p, such that $(p_*)^+ = p_*^+$. Further, p_* is unlimited, and if $A = M$ is a p.i. model, then there is some $\overline{u} \in U \models T_U$ and some $\overline{m}$ in M such that $(M \oplus U > M$ and) $p = t(\overline{m}+\overline{u}/M)$ and $p_* = t(\overline{u})$.

Proof: Only the latter needs a proof. Let $N > M$ realize p. As M is p.i., there is some U as above with $N = M \oplus U$ and also $\overline{m}+\overline{u}$ realizing p as above. If $\phi(\overline{x}) \,\varepsilon\, p_*^+$ then $\phi(\overline{x}) = \phi(\overline{x},\overline{0})$ for some $\phi(\overline{x},\overline{m}') \,\varepsilon\, p^+$. Then $\models \phi(\overline{m}+\overline{u},\overline{m}')$ implies $\models \phi(\overline{u},\overline{0})$. Thus $p_*^+ \subseteq t^+(\overline{u})\ (= t_*^+(\overline{u}))$.

For the converse inclusion let $\phi(\overline{u}) \varepsilon t^+(\overline{u})$. Put $\chi(\overline{x},\overline{y}) = \phi(\overline{x}\text{-}\overline{y})$. Then $\chi(\overline{x},\overline{m}) \varepsilon p^+$, hence $\phi \varepsilon \overset{+}{p_*}$. □

Lemma 5 (Pillay and Prest). If p is in S(A) and $B \supseteq A$ then there is $p \subseteq q \in S(B)$ with $p_* = q_*$.

Proof: I have to show that $p \cup \{\neg\psi(\overline{x},\overline{b})\text{: } \overline{b} \varepsilon B \text{ and } \psi(\overline{x},\overline{0}) \notin \overset{+}{p_*}\}$ is consistent. If not, there are finitely many $\phi(\overline{x},\overline{a}) \varepsilon p^+$, $\phi_i(\overline{x},\overline{a}) \varepsilon p^-$, $\psi_j(\overline{x},\overline{0}) \notin \overset{+}{p_*}$ and $\overline{b}_j \varepsilon B$ such that $\phi(\overline{x},\overline{a})$ is contained in the union of the $\phi_i(\overline{x},\overline{a})$ and the $\psi_j(\overline{x},\overline{b}_j)$. Since p is consistent, at least one of the ψ_j must occur, ψ_0 say. On the other hand, BHN allows us to omit those for which $\phi/\phi\wedge\psi_j$ is infinite. Thus $\phi/\phi\wedge\psi_0$ is finite, hence $\psi_0 \varepsilon \overset{+}{p_*}$; contradiction. □

Notice that in general $t_*(\overline{c}/A) = t_*(\overline{a}+\overline{c}/A)$ for all $\overline{a} \in A$.

Actually I am working in two theories, T and T_U, and it can therefore happen that some $p \in S^T(0)$ and $q \in S^{T_U}(0)$ are different even though $p^+ = q^+$ (for $T \subseteq p$ and $T_U \subseteq q$). Also a little accuracy is needed when talking about pp types, since even being a filter depends on the theory (look at q containing $px=0 \wedge x\neq 0$ in $T = Th(\mathbb{Z}(p^\infty) \oplus Q)$; there q^+ is a filter, whereas in T_U, which is Th(Q) in this case, every pp type containing px=0 contains also x=0).

Nevertheless, the next fact, taken from [B-R], justifies any confusion of T and T_U, at least when dealing with unlimited types.

Lemma 6. (1) $p \mapsto p_*$ defines a surjective map from S(A) onto $S_*(0)$.

(2) There is a bijection between $S_*(0)$ and $S^{T_U}(0)$ bringing every $q \in S_*(0)$ to some $p \in S^{T_U}(0)$ so that $p^+ = q^+$.

(3) Let $q \in S(0)$ be realized in some $M \oplus U > M \models T$ (i.e. $U \models T_U$). Then q is realized by some tuple of U iff $q \in S_*(0)$.

Proof:

(1) The preceding lemma allows us to assume A = 0. Then, however, the assertion is trivial, since $p_* = p$ for all $p \in S_*(0)$.

(2) If $q \in S_*(0)$ then - according to Lemma 3(2) - choose $p \in S^T U(0)$ with $p^+ = q^+$. If $p \in S^T U(0)$ then - using (1) of the same lemma - choose q in $S(0)$ so that q^+ is the closure of p^+ under pp subgroups of finite index. Clearly, $q_* = q$, hence $q \in S_*(0)$.

(3) If $q = t(\bar{u})$ for some $\bar{u} \in U$ then $q = q_* \in S_*(0)$ by Lemma 1. If $q \in S_*(0)$, choose $q \subseteq p \in S(M)$ with $q_* \ (= q) = p_*$. Write $p = t(\bar{m}+\bar{u}/M)$, where $\bar{m} \in M$ and $\bar{u} \in U$. As in Corollary 4, $q = p_* = t(\bar{u})$. □

Let $\underline{S_*^+(0)}$ denote the set $\{p^+ : p \in S_*(0)\}$. From Lemma 3 it follows that $S_*^+(0)$ is closed under arbitrary unions, and contains the set of all pp subgroups of finite index and clearly also the set of all pp subgroups. The 1-types from $S_*(0)$ having these latter as pp part are denoted by **1** and **0**, correspondingly. (From now on all types are 1-types, similarly for sets of those!) There is another way of looking at $S_*^+(0)$. Let U be a model of T_U realizing all types over 0. Then every type $p \in S_*^+(0)$ corresponds to some (infinitely definable) subgroup $p(U)$ of U. Then accordingly $\{p(U) : p \in S_*^+(0)\}$ forms a complete modular lattice under $\cap$ and + (cf. also [PR, §8.11]). $U = \mathbf{1}^+(U)$ is the greatest and $0 = \mathbf{0}^+(U)$ the smallest element in this lattice. Since this lattice does not depend on U (as long as U realizes all relevant types), I will denote it by $\underline{pp_\infty(T_U)}$. It is also called the <u>lattice of infinitely pp definable subgroups of T_U</u>. If T_U has dcc on pp subgroups then $pp_\infty(T_U)$ reduces merely to the lattice $pp(T_U)$ of pp subgroups of (some (=any) model of) T_U.

I will need also the following description of algebraic types, which appeared in the proof of [Ro 1, Lemma 2].

Lemma 7. A complete type is algebraic iff it contains a coset of a finite pp subgroup.

Proof: For the non-trivial direction, let $\phi \wedge \bigwedge_{i<n} \neg \psi_i$ be non-empty but finite, where $\phi \in p^+$, $\psi_i \in p^-$. Then there are finitely many elements $a_0,\dots,a_{n-1}$ $(n>0)$ such that $\phi \subseteq \bigcup_{i<n} \psi_i \cup \bigcup_{j<n} \{a_j\}$ and no a_j can be omitted. By BHN, $\phi(x,\bar{0})/0$ is finite, for $\{a_j\}$ is a coset of the trivial group 0. □

Corollary 8. A complete type is algebraic iff $p_* = \mathbf{0}$. □

The finitizer $fin_R M$ of an R-module M is, by definition, the set of ring elements r such that rM is finite ([Ro 3]). Let us sum up some useful properties of the finitizer (cf. ibid.).

Lemma 9. Let $I = fin_R M$.

(1) I is an ideal containing the annihilator $ann_R M$.

(2) I does not depend on $M \models T$ (also neither does $ann_R M$). So the notations fin T and ann T make sense.

(3) $I = \{r \in R: M/M[r] \text{ is finite}\}$, where $M \models T$ and, as usual, $M[r]$ is the pp subgroup of M defined by $rx=0$.

(4) fin $T = $ ann T_U.

(5) $IM \subseteq$ acl 0, whence the factor module M/acl 0 is an R/I-module (here, as common in model theory, acl denotes algebraic closure; it is easily seen that acl 0, which is the same as acl $\varnothing$ and which is the same in each model of T, is a submodule of every model).

Proof:

(1) Let rM and r'M be finite. Then $(r+r')M = rM + r'M$ and s(rM) and r(sM) must be finite, too. (2): "$|rM| = n$" is a first-order statement. (3): For every $r \in R$ consider the endomorphism h_r of the additive group of the model M of T. Then Ker $h_r = M[r]$ and im $h_r = rM$. Thus $M/M[r] \simeq rM$. (4) follows from (3) and Lemma 1. (5) is clear. □

§2. Forking.

The aim of this section is to introduce, mostly without proof, module-theoretic equivalents to stability-theoretic concepts as forking, Lascar rank, and regular type.

Again I will work in a fixed completion T of T_R in this section. Let p and q be complete types. q is a nonforking (or nf-) extension of p, in terms $p \sqsubset q$ or $q \sqsupset p$, if $p \subseteq q$ and $p_* = q_*$. If $p \in S(A)$ then we also say, q does not fork over A (or q dnf/A). q is a forking extension of p, in terms $p \not\sqsubset q$ or $q \not\sqsupset p$, if $q \supseteq p$, but q is not an nf-extension of p. That this coincides with the usual notion due to Shelah was shown in [Ga] for theories closed under products (see also [Zie]) and in full generality in [P-P 1].

Lascar's U-rank is an ordinal rank on complete types, which is defined as follows (see e.g. [SH], where it is called L):

(a) $RU(q) \geq 0$ for all complete q;

(b) $RU(q) \geq \delta$, where δ is a limit ordinal, if $RU(q) \geq \alpha$ for all $\alpha < \delta$;

(c) $RU(q) \geq \alpha+1$ if there is a complete $r \not\sqsupset q$ with $RU(r) \geq \alpha$;

(d) $RU(q) = \alpha$ if $RU(q) \geq \alpha$ and $RU(q) \not\geq \alpha+1$;

(e) $RU(q) = \infty$ if $RU(q) \geq \alpha$ for any ordinal α.

As noticed in [B-R], the above definition of forking together with Lemma 1.6 shows that, for a complete type p in T, RU(p) is just the foundation rank of p_*^+ in $pp_\infty(T_U)$ (with respect to the order given by inclusion), i.e. $RU(p) = RU(p_*)$, and on $S_*(0)$ we have

(a) $RU(q) \geq 0$ for all $q \in S_*(0)$;

(b), (d), and (e) as above;

(c) $RU(q) \geq \alpha+1$ if there is an $r \in S_*(0)$ such that $RU(r) \geq \alpha$ and $q^+ \subset r^+$ (or, equivalently, $q^+(U) \supset r^+(U)$ in $pp_\infty(T_U)$). Clearly, $RU(\mathbf{O}) = 0$ and, moreover, $RU(q) = 0$ iff $q = \mathbf{O}$ (for $q \in S_*(0)$). Thus, by Corollary 8, $RU(p) = 0$ iff p is algebraic (for arbitrary complete p; this is a general fact, cf. [SH]). Further, if $p_* = \mathbf{1}$ then $RU(p) \geq RU(q)$ for all complete types q.

Forking can be defined in any stable theory ([SH]; modules are stable as shown by Baur and Fisher). Using forking, in turn, one can define a notion of independence as follows: The sets B and C are independent over a set A, in terms $B \downarrow_A C$, if $t(B/A \cup C)$ dnf/A. This kind of independence relation has a number of nice properties, however, the corresponding relation of dependence lacks one of van der Waerden's axioms: it is not transitive. Therefore Shelah introduced regular types and proved that forking dependence is in fact transitive on elements realizing regular types. Based on that he developed an elaborate dimension theory for arbitrary stable structures.

Now, a type $p \in S(A)$ is called regular if for all $B \supseteq A$ and a,b realizing p, if $a \downarrow_A B$ and $b \not\downarrow_A B$ then $a \downarrow_B b$.

I conclude this section with a module-theoretic description of this, which was obtained in [Zie] for theories closed under products and in full generality in [Pr].

Fact 1 For a complete type p in T the following are equivalent:

(1) p is regular.

(2) p_* is regular.

(3) p_* is critical in the sense that p_*^+ defines a minimal nonzero infinitely definable pp subgroup in $H(p_*)$ (in other words, p_*^+ is a maximal nonzero pp type in $Th(H(p_*))$.

Further, if p is regular, p_* is indecomposable.

§3. Stability.

As mentioned in the preceding section, all complete theories of modules are stable (it is this what we mean when we say that all modules are stable). Two important subclasses of that of stable theories appear in stability theory. I will define them only in the context of modules. That those definitions coincide with the original ones (given in terms of the power of certain Stone spaces) is due to Garavaglia (and depends on Baur's pp quantifier elimination for modules, see e.g. [Zie] for proofs and any further details concerning this section). Fix a completion T of T_R again. Remember, "types" are "1-types" now.

T is totally transcendental (t.t.) if it - or rather each (= some) of its models - satisfies the descending chain condition (dcc) on pp subgroups. T is superstable (s.s) if it satisfies the weak dcc on pp subgroups, i.e. no model contains an infinite chain of pp subgroups in which every factor is infinite.

Lemma 1 [B-R, 3.2]. T is s.s. iff T_U is t.t.

Proof: In any chain of pp subgroups of some $M \models T$, by Lemma 1.1, the infinite factors do not collapse in T_U. Hence the dcc for T_U implies the weak dcc for T.

Conversely, let $\phi_0(U) \supset \phi_1(U) \supset \phi_2(U) \supset \ldots$ be an infinite chain of pp subgroups in some $U \models T_U$. Consider $\psi_0 = \phi_0$, $\psi_{i+1} = \psi_i \wedge \phi_{i+1}$. In U the ψ_i define the same chain, and they also define a chain in every other module. But in a model M of T, in addition, the factors $\psi_i(M)/\psi_{i+1}(M)$ are infinite, by Lemma 1.1, as $M \oplus U \models T$. □

Corollary 2. A theory closed under products is s.s. iff it is t.t. □

Lascar proved the next result for arbitrary theories. In case of modules, though, it is particularly easy to prove.

Corollary 3. T is s.s. iff every complete type has an ordinal U-rank.

Proof: T is s.s. iff T_U is t.t. iff $pp_\infty(T_U) = pp(T_U)$ iff $pp_\infty(T_U)$ is well-founded iff each $p \in S_*(0)$ has an ordinal foundation rank in $pp_\infty(T_U)$. □

Let us return to t.t. theories again. If T is t.t. then, by the dcc on pp subgroups (which immediately yields the dcc on their cosets), every pp type (even with parameters) is equivalent to a single pp formula. We call such types finitely generated (f.g.). (In general, following Prest we call a complete type finitely generated if its pp part is.) From this it is not hard to derive that every model M of T is p.i., even Σ-p.i. - which means that $M^{(\omega)}$ is also p.i., for M satisfies the dcc iff $M^{(\omega)}$ does (since $\phi(M^{(\omega)}) = [\phi(M)]^{(\omega)}$ for any pp formula $\phi(x)$).

This makes the classification of models of t.t. theories of modules particularly transparent. For instance, Garavaglia used this to show that a countable t.t. theory of modules satisfies Vaught's conjecture, i.e. such a theory has either countably or continuum many isomorphism types of countable models (see also the cited literature).

That every model be p.i. can be violated even in s.s. theories (as a matter of fact, no countable module with an infinite descending chain of pp subgroups can be p.i., as there would be too many pp types to be realized). Therefore it is often very important to know whether a certain theory is t.t. or not. A much more powerful criterion for total transcendence than that of all types being finitely generated is the following, which I state without proof.

Fact 4 [P-P 2, 6.8]. T is t.t. iff every regular unlimited type over 0 is finitely generated.

Note, this implies that for total transcendence it is enough to check indecomposable unlimited types (Fact 2.1) - in T, though, for a type p in $S_*(0)$ can, in general, have finitely generated pp part when considered in T_U without being finitely generated in T.

§4. U-rank 1 modules.

From a stability-theoretic point of view the easiest theories to consider are those in which every complete type has U-rank at most 1. Theories having this property (and even their models) are also said to have U-rank 1. These are also characterized by the fact that no non-algebraic 1-type forks over the empty set. In modules we have

Lemma 1 [B-R, Corollary 2], [P-P 2, Proposition 7.1].

The following are equivalent for any completion T of T_R:

(1) T has U-rank 1.

(2) (Every (=some) model of) T_U is pp-simple, i.e. there are no proper pp subgroups in (models of) T_U.

(3) Every pp subgroup in T is either finite or has finite index in (a model of) T.

Proof: From §2 we know that T has U-rank 1 iff $pp_\infty(T_U) = \{\mathbf{1}^+, \mathbf{0}^+\}$. Thus the lemma follows from Lemma 1.1. □

Corollary 2 [P-P 2]. If T has U-rank 1 then there is exactly one nonzero type in $S_*(0)$. This type is regular and hence indecomposable.

Proof: The first statement is immediate using Lemma 1.6(2). Further, a pp-simple module is p.i., as there are only trivial pp types, which are realized anyway. Thus the p.i. hull of a realization $u \in U \models T_U$ of the unique unlimited type of T is contained (as a direct summand) in U, consequently pp-simple itself. Then that type is critical. It remains to apply Fact 2.1. □

Corollary 3. If $F = R/\mathrm{fin}_R M$ is a skew-field for some (=any) $M \models T$, then T has U-rank 1.

Proof: By Lemma 1.9(4), every $U \models T_U$ is an F-module, hence a vector space over F. Vector spaces are pp-simple, since every pp formula defines a right ideal in $_RF$ (easily checked). □

§5. Regular generics.

Proceeding from the U-rank 1 case to more complicated cases by successively allowing the U-rank grow, one quickly gets into trouble caused by the fact that pp subgroups need not be submodules if the ring is not commutative. Poizat, however, introduced a class of theories containing properly that of U-rank 1 theories, which retains some of the latter's nice properties. These are the theories which have regular generic types, where, following Poizat, a generic type over some model M is a type $t(c/M)$ such that $t(c+m/M)$ dnf/0 for all $m \in M$. An arbitrary complete type is called generic if its non-forking extensions over some model are generic. Notice, generic types do not fork over 0.

Again, the context is a fixed completion T of T_R. Most of what is contained in this section was announced in [Ro 2].

Lemma 1. A complete type p is generic iff its free part p_* is **1**.

Proof: Using Corollary 1.4 write $p_* = t(u)$ for some $u \in U$ such that

$c = m' + u \in M \oplus U > M$. As mentioned after Lemma 1.5, $t_*(c+m/M) = t(u)$ for all m in M. As $p = t(c/M)$ is generic iff all $t(c+m/M)$ dnf/0, we have that p is generic iff all $t_*(u+m) = t(u)$. This is equivalent to $t_*^+(u+m) = t^+(u)$ for all $m \in M$.

I claim, $t_*^+(u+m) = t^+(u) \cap t_*^+(m)$.

For the inclusion from left to right, let $\phi \in t^+(u+m)$ and ϕ/ψ be finite. Then $\phi \in t^+(u)$ and $\phi \in t^+(m)$. Thus $\psi \in t^+(u)$ and $\psi \in t_*^+(m)$, since $\phi(U) = \psi(U)$. The other inclusion is left to the reader.

Consequently, p is generic iff $t^+(u) = t^+(u) \cap t_*^+(m)$ for all $m \in M$ iff $t^+(u) \subseteq t_*^+(m)$ for all m iff $t^+(u)$ contains no pp subgroup of infinite index of M iff $t(u) = \mathbf{1}$. □

Thus there is a distinguished generic type, namely **1**, and when talking of <u>the</u> generic we always mean this type.

Together with the description of regularity from §2 we get that all generics are regular iff some is (this is a general fact, see [Po]) iff **1** is.

Lemma 2. The following are equivalent:

(1) T has regular generics.

(2) **1** is regular.

(3) H(**1**) is pp-simple.

(4) There is an element a in some $U \models T_U$ such that H(a) is pp-simple and a lies in no proper pp subgroup of U (i.e. every such intersects H(a) trivially).

Proof: The equivalence of (1) - (3) has been mentioned already. For the last one notice that such an element a has a generic type over 0 iff t(a) is **1** (in T). □

Part of the corollary is contained also in [Po, Lemme 7], (4) is a special case of [Po, Lemme 2].

Corollary 3. Let T have regular generics and put $\bar{R} = R/\text{fin } T$.

(1) $H(\mathbf{1})$ is indecomposable.

(2) $\phi(H(\mathbf{1})) \neq 0$ iff $\phi(H(\mathbf{1})) = H(\mathbf{1})$ iff $U = \phi(U)$ for all $U \models T_U$ iff $M/\phi(M)$ is finite for all (some) $M \models T$ (ϕ pp).

(3) $\mathrm{ann}_R H(\mathbf{1}) = \mathrm{ann}\ T_U = \mathrm{fin}\ T$.

(4) rM is either finite or of finite index in M, for each $M \models T$, $r \in R$.

(5) $H(\mathbf{1})$ is a torsion-free divisible $\bar{R}$-module.

(6) Every $U \models T_U$ is a divisible $\bar{R}$-module.

(7) fin T is a prime ideal, i.e. $\bar{R}$ a domain.

Proof:

(1) follows from Fact 2.1.

(2) is immediate from Lemma 2 and Lemma 1.1.

(3) Applying (2) to the formula $rx=0$ we get $\mathrm{ann}\ T_U = \mathrm{ann}_R H(\mathbf{1})$. The other equality was proved in §1.

(4) Applying (2) to the formula $\exists y(x=ry)$ we get, M/rM is finite iff $rH(\mathbf{1}) \neq 0$. This is the case iff $r \notin$ fin T iff rM is infinite.

(5) and (6) As in (4) one easily gets $rH(\mathbf{1}) = H(\mathbf{1})$ - and by (2), $rU = U$ - whenever $r \notin$ fin T. Hence $H(\mathbf{1})$ and U are divisible (over $\bar{R}$).

A module is torsion-free as an $\bar{R}$-module iff $rx=0$ defines in it the trivial group 0 for all $r \notin$ fin T.

If $rx=0$ defines a non-trivial group in $H(\mathbf{1})$ then it defines the whole group, by pp-simplicity. But then $r \in$ fin T.

(7) easily follows from (3) and Lemma 2 (3). □

PART II. THE CONNECTED CASE.

Modules with regular generics will be investigated here in the case when they do not contain proper definable subgroups of finite index. These turn out to be divisible and, essentially, over rings without zero divisors. Most of this part is devoted to the search for a sufficiently large class of domains over which every divisible module has regular generics.

§6. Connected modules.

A module is called <u>connected</u> if it does not contain any proper definable subgroup of finite index.

A module is connected iff the pp part of any generic over 0 is just $\{x=x\}$:

Lemma 1. A module is connected iff it does not contain any proper pp subgroup of finite index.

Proof: For the non-trivial direction, suppose all proper pp subgroups of a module M have infinite index and let A be a definable subgroup of finite index. Then there are finitely many pp formulas ϕ_i, ψ_{ij} so that $\bigcup_i (\phi_i \wedge \bigcup_j \neg\psi_{ij})$ defines A in M. Let there be no redundant disjunct. As A has finite index, M is a finite union of cosets of the ϕ_j. By BHN there is one, ϕ_o say, which has finite index in M. By hypothesis, $\phi_o(M)$ is just M. By irredundancy, the first disjunct is not empty in M, hence all the ψ_{oj} are proper and thus of infinite index. On the other hand, $M = (M \setminus \bigcup_j \psi_{oj}) \cup \bigcup_j \psi_{oj}$. By BHN, we can omit the ψ_{oj}, whence $M \subseteq A$. □

Lemma 2. If M is connected, $fin_R M = ann_R M$.

Proof: By Lemma 1.9(3), $fin_R M = \{r \in R : M/M[r] \text{ is finite}\} = \{r \in R : M = M[r]\} = ann_R M$. □

Notice that also connectedness is an invariant of a complete theory, as is divisibility. Thus it makes sense to talk about <u>divisible</u> or <u>connected</u> complete theories. I might admit this sort of confusion also in other cases of first-order invariant properties.

Throughout, T is a completion of T_R.

Corollary 3. If T has regular generics and is connected, then $rM = M$ for each $M \models T$ and $r \in R \setminus ann_R M$.

In particular, the ring $\bar{R} = R/ann_R M$ is a domain and M a divisible $\bar{R}$-module.

Proof: Follows from the preceding lemma and Corollary 5.3. □

Corollary 4. Let $\bar{R}$, T, M be as above. Suppose $\bar{R}$ is finite. T has regular generics and is connected iff $\bar{R}$ is a field. M is an $\bar{R}$-vector space then, hence pp-simple, and T is t.t.

Proof: Finite rings without zero divisors are skew-fields, hence fields. If, on the other hand, $\bar{R}$ is a field, then, being a vector space, M is pp-simple, hence connected and t.t., and also p.i. Thus H(**1**) is $\bar{R}$, hence pp-simple. □

I am particularly interested in whether a given module with regular generics is t.t. Even in the connected case there is no hope of showing all of them are:

Example: Let $R = k[X,Y]$, the ring of polynomials over a field k in two commuting indeterminates X and Y. Let Q be its field of fractions $k(X,Y)$. Consider the R-module $M = Q/R$. That M is in fact a connected module with regular generics follows from the next section (any divisible module over a commutative domain is). As in every divisible torsion module, there are plenty of ascending chains of the form $M[r] \subset M[r^2] \subset \ldots$ However, I am going to find also a descending chain of pp subgroups (this making M non-t.t.). Clearly, $M[X]$, $YM[X]$, $Y^2M[X]$,... are pp subgroups, and since X and Y commute, they form a descending chain. It remains to show that this is proper. For this it is enough to verify that $Y^n/X + R$ is not in $Y^{n+1}M[X]$, for all n. Otherwise write $Y^n/X + R = Y^{n+1}(r/s + R)$, where $X(r/s + R) = 0$, i.e. $Xr =$

su for some $u \in R$. From the former we get some $v \in R$ with $Y^n/X = (Y^{n+1}r)/s + v$, hence $Y^n = Y^{n+1}u + Xv \in XR + Y^{n+1}R$, contradiction. □

This shows that we have to study divisible modules in further detail. I will be interested in which of these have regular generics and which of those, in turn, are t.t. Unfortunately I do not have a full answer to either of these questions. Eventually I will show though that every divisible module has regular generics and is connected if we restrict to rings which are e.g. two-sided Ore domains. This is not too much a restriction, for it includes the commutative case and the two-sided noetherian case (see below).

§7. Divisible modules.

Before turning to the results mentioned at the end of the last section I will derive some preliminary facts about pp subgroups in divisible modules which do not depend on further restrictions on the ring. Nevertheless I confine myself to domains, since the main interest is in modules with regular generics anyway.

A large subclass of that of divisible modules is that of injectives, for divisibility is equivalent to Baer's criterion restricted to principal ideals. These can differ quite a bit: e.g. over left-noetherian rings any injective module is t.t. (see cited literature), whereas the above example provides a divisible one which is not.

Recall from Robinson style model theory that a formula over a module N (i.e. with parameters from N) is <u>model consistent</u> with N if it is satisfiable in some supermodule of N (i.e. iff it is consistent, in the usual sense, with the atomic diagram of N). I will consider this notion for finite systems of equations only. For this purpose I fix some

<u>Notation</u>: For a pp formula $\phi(x)$ of the form $\exists\overline{y}\,\psi(x,\overline{y})$, where $\psi(x,\overline{y})$ is $\bigwedge_{i<n} s_i x = \sum_{j<m} s_{ij}y_j$ (and clearly $\ell(\overline{y}) = m$), put $\overline{s}_i = (s_{i0},\ldots,s_{i(m-1)}) \in R^m$ and let I_ϕ be the submodule of ${}_RR^m$ generated by

$\bar{s}_0,\ldots,\bar{s}_{n-1}$. Furthermore, let $h_\phi : {}_RR^n \to {}_RR^m$ be the homomorphism given by $(r_0,\ldots,r_{n-1}) \to (\sum_{i<n} r_is_{ij} : j < m)$ $(= \sum_{i<n} r_i\bar{s}_i)$. □

Next I state Lemma 3.2 from [E-S] specified to this notation (notice, the restriction to one free variable in [E-S] is not essential).

Fact 1. For $\phi(x)$ as above, a module N and some $a \in N$ the following are equivalent (in the above notation):

(1) $\psi(a,\bar{y})$ is model consistent with N;

(2) $\sum_{i<n} r_is_ia = 0$ for all $r_i \in R$ such that $\sum_{i<n} r_i\bar{s}_i = 0$ (i.e. for all $(r_0,\ldots,r_{n-1}) \in \mathrm{Ker}\ h_\phi$);

(3) there is a homomorphism $g : I_\phi \to N$ with $g(\bar{s}_i) = s_ia$ for all $i < n$.

(The proof of (3) → (1) uses some amalgamated sum of N^m and ${}_RR^m$ over I_ϕ, the other implications are trivial). □

Call a submodule $M \subseteq N$ <u>n-pure</u> in N if for each n-place pp formula $\phi(\bar{x})$ and each $\bar{a} \in M^n$, $M \models \phi(\bar{a})$ iff $N \models \phi(\bar{a})$.

M is called <u>absolutely n-pure</u> if it is n-pure in every extension. The usual notion of purity is that where n is allowed to run over all natural numbers. Similarly for absolute purity. Actually I am interested only in the case $n = 1$. Absolute 1-purity is a notion between divisibility and injectivity (for a direct summand is clearly n-pure) as is the $\aleph_0$-injectivity of [E-S], which is somehow oblique to absolute 1-purity, though. It is easily seen that N is absolutely 1-pure iff every system of equations of the form $\{s_ia = \sum_{j<n} s_{ij}y_j : i < n\}$, where $a \in N$, has a solution in N if it is model consistent with N.

Lemma 2. For an arbitrary pp formula ϕ the following are equivalent (in the above notation):

(1) $\phi(N) = N$ for all absolutely 1-pure (left R-) modules N;

(2) $\phi(M) = M$ for all injective modules M;

(3) $\psi(a,\bar{y})$ is model consistent with N for every module N and each $a \in N$;

(4) $\psi(1,\bar{y})$ is model consistent with ${}_RR$;

(5) $(r_0,\ldots,r_{n-1}) \in \text{Ker } h_\phi \Rightarrow \sum_{i<n} r_i s_i = 0.$

If these conditions are violated then there is an $r \in R^o$ such that $\phi(M) \subseteq M[r]$, in particular, $\phi(M) \subseteq T(M)$, where T(M), the torsion part of M, is the set $\{a \in M : ra=0$ for some $r \in R^o\}$, for every module M.

If there is some torsion-free injective $N \neq 0$ with $\phi(N) = N$ then this is true for every injective module N.

Proof:

(1) → (2): Each injective is absolutely pure.

(2) → (3): Consider an injective $M \supseteq N$ (e.g. its injective hull). Then $M \models \phi(a)$, hence $\psi(a,\bar{y})$ is realized in M.

(3) → (4): is trivial, (4) → (5) → (3) is the above fact.

(3) → (1): Let N be absolutely 1-pure and $a \in N$. Choose $M \supseteq N$ such that $M \models \exists \bar{y}\, \psi(a,\bar{y})$. As N is 1-pure in M, $N \models \phi(a)$.

Thus (1) through (5) are equivalent.

Next assume, (5) is violated. Then there are $r_i \in R$ $(i < n)$ with $r = \sum_{i<n} r_i s_i \neq 0$, but $\sum_{i<n} r_i s_{ij} = 0$ for all $j < m$, hence $\sum_{i<n} r_i(s_i a) = 0$ if $a \in \phi(M)$ (M arbitrary). Then $ra=0$, whereby the second assertion follows. Thus, if there is a torsion-free N as above, the conditions (1) - (5) must hold.

□

Corollary 3. Over domains, torsion-free absolutely 1-pure modules are pp-simple, and thus t.t.; torsion-free absolutely pure modules are injective.

Proof: A proper pp subgroup of an absolutely 1-pure module is in the torsion part. An absolutely pure p.i. module is injective. □

Lemma 4. If R is a domain, then (-1) → (0) → (1) for any pp formula $\phi(x)$, where

(-1) $\phi({}_R R) \neq 0$;

(0) $\phi(M) = M$ for all divisible (left R-) modules M;

(1) $\phi(N) = N$ for all absolutely 1-pure modules N (as in Lemma 2).

Proof:

As absolutely 1-pure modules are divisible, only (-1) $\rightarrow$ (0) needs a proof. Let $\phi(x)$ be $\exists\bar{y} \bigwedge_{i<n} (s_i x = \sum_{j<m} s_{ij} y_j)$. If $0 \neq r \in \phi({}_RR)$, there are $r_j \in R$ such that $s_i r = \sum_{j<m} s_{ij} r_j$ $(i < n)$. If M is divisible and $a \in M$ arbitrary, pick $b \in M$ with $a = rb$ (here we need $r \neq 0$!). Then $s_i a = s_i(rb) = \sum_{j<m} s_{ij}(r_j b)$, whence $a \in \phi(M)$. □

For the rest of this part I will be dealing with rings for which some of the converses of these implications are true. Namely, domains which are characterized by (1) $\rightarrow$ (0) I will call good in the next section, and in the appendix I will show that right Ore domains are exactly the domains satisfying (0) $\rightarrow$ (-1).

§8. Modules over Ore domains.

I am going to single out a large enough class of domains, over which every divisible module has regular generics (and is connected). This provides us with a stock of examples automatically including those which are not injective (and, as announced in [Ro 2], not even absolutely pure).

To realize this task I impose two further restrictions on the domain, the classical left Ore condition (which is (a) in Lemma 2 below) and the following less classical one.

A ring is <u>good</u> if the following holds for all 1-place pp formulas $\phi(x)$: If $\phi(M) = M$ for all injective M then $\phi(M) = M$ for all divisible M. (Notice also the equivalent formulations of this given by Lemma 7.2). Good domains are less exotic than their definition might let them seem to be. As a matter of fact, every right and left Ore domain is good, as I will show in the appendix. In particular, commutative domains and also right and left noetherian domains are good (cf. [ST, Ch.II] for the fact that noetherian rings are Ore). However, I do not know whether every good left Ore domain is right Ore.
Goodness guarantees plenty of generic elements:

Lemma 1. Every torsion-free element (i.e. everyone not in T(M)) in a divisible module M over a good ring realizes a generic type.

Proof: Let M be divisible and consider an arbitrary pp subgroup $A \neq M$. Since the ring is good, Lemma 7.2 implies $A \subseteq T(M)$. Hence no element outside T(M) lies in a proper pp subgroup. □

The other property I impose on the ring will make sure that the generics are regular. It is introduced - among others - in the next lemma, which is basically folklore. For completeness I enclose a proof.

Lemma 2. Let R be a domain.

(1) The following are equivalent:

(a) $Rs \cap Rr \neq 0$ for all r and s from R^o;

(b) T(M) is a submodule in every (left R-) module M;

(c) T(M) is closed under scalar multiplication in every module M.

A domain having these properties is called <u>left Ore</u>.

(2) T(M) is <u>divisor closed</u>, i.e. $Ra \cap T(M) = 0$ for all $a \in M \setminus T(M)$. In particular, T(M) is divisible if it is a module.

(3) If R is left Ore then M/T(M) is a torsion-free R-module for every (left R-) module M.

(4) If R is left Ore then torsion-free divisible (left R-) modules are injective (hence pp-simple and t.t. by Corollary 7.3).

Proof: (1) (a) →(b): For $a_i \in T(M)$ choose $s_i \in R^o$ with $s_ia_i = 0$ $(i < 2)$. Also choose $r = r_os_o = r_1s_1 \neq 0$ using (a). Then $r(a_o+a_1) = 0$, whence T(M) is a subgroup. If a_o,s_o are as before and $s_1 \in R^o$ is arbitrary, pick r_o and r_1 as before. Then $r_1s_1a_o= 0$, hence $s_1a_o \in T(M)$. (b) follows. (b) → (c) is trivial. (c) → (a): Fix $s_1 \in R^o$. I will show that for all s_o in R^o there is an $r \in R^o$ with $rs_o \in Rs_1$. This, however, is nothing else than T(N) = N for the module $N = R/Rs_1$. As $s_1.1 \in Rs_1$, $1+Rs_1 \in T(N)$. Then, by (c), T(N) contains every $r+Rs_1$. This completes the proof of (1). (2) just uses that R is a domain. (3) follows from (1) and (2). (4): I am going to verify Baer's criterion for injectivity, which requires finding b in M for every non-zero left ideal J and every homomorphism h from J into M such that $h(r) = rb$ for all r in J.

Pick any non-zero t in J and divide h(t) by t, i.e. find a b in M with $h(t) = tb$. Given an arbitrary non-zero r in J, choose $sr = s't \neq 0$ in R using Ore's condition. Then $sh(r) = h(s't) = s'(tb) = s(rb)$. Torsion-freeness yields $h(r) = rb$. □

The result I am heading for is now

Theorem: Let M be a left module over a good left Ore domain R. M is divisible iff it has regular generics and is faithful and connected.

One half of the theorem is a special case of Corollary 6.3. I am going to prove the other in a number of steps. Remember that because of Corollary 6.4 I need not consider finite rings.

Let me point out that assuming $ann_R M$ to be 0 in the theorem is of no substantial loss, for we could just work with $\bar{R}$ instead (the annihilator being an invariant of T).

Lemma 3. A faithful module over a left Ore domain has an elementary extension which is not torsion.

Proof: If the torsion-free type $\{rx \neq 0 : r \in R^o\}$ is inconsistent with M, then $M \models \forall x \bigvee_{i<n} r_i x = 0$ for some $r_0, \ldots, r_{n-1}$ in R^o, by compactness. As R is left Ore, there is $0 \neq s \in \bigcap_{i<n} Rr_i$. Then $sM = 0$, contradicting faithfulness. □

Recall that a divisible module over a domain is certainly faithful. I call a module M non-torsion if it is not a torsion module i.e. iff $T(M) \neq M$.

Lemma 4. Every divisible non-torsion module over an infinite good domain is connected.

Proof: As in Lemma 1, a proper pp subgroup $\varphi(M)$ is contained in T(M). We will show that $M/\varphi(M)$ is infinite if it is non-zero. Pick an element a outside T(M). By Lemma 2(2), $Ra \cap \varphi(M) = 0$. Thus all the ra, where r runs over R, are in distinct cosets of $\varphi(M)$. The lemma now follows from the infiniteness of R. □

The next lemma completes the proof of the theorem.

Lemma 5. If M is a divisible module over an infinite good left Ore domain then $M \equiv M \oplus (M/T(M))$.

If, in addition, $M \neq T(M)$ then the generics are regular.

Proof: It suffices to show that $\phi/\psi(M/T(M)) = 0$ whenever $\phi/\psi(M)$ is finite. Assume, the former is non-zero. Then, using the pp-simplicity of M/T(M) established in Lemma 2(4), we get $\phi(M/T(M) = M/T(M)$ and $\psi(M/T(M)) = 0$. Now we apply Lemma 7.2: "Since ψ is 0 in the injective M/T(M), it is inside the torsion part in every module. In particular, $\psi(M) \subseteq T(M)$. Further, since φ defines the whole module in the torsion-free injective M/T(M), it does so in all divisible modules (R being good!). In particular, $\phi(M) = M$. Consequently, $\phi/\psi(M)$ is infinite by the bracketed statement of the preceding lemma. This completes the proof of the first assertion and also shows that a and a+T(M) have the same type if $a \in M \setminus T(M)$.

The second assertion can be derived from the fact that every element a outside T/(M) is generic and has, moreover, pp-simple p.i. hull: Namely, M/T(M) is p.i. (even injective), hence it contains the hull of each of its elements, which have therefore pp-simple hulls, too. □

The theorem is proved.

Let me finally return to the example in §7.

The non-t.t. module M considered there is divisible. The ring R is a commutative domain, hence left and right Ore. By what will be shown in the appendix those are good. Consequently, M has regular generics.

Appendix. Two-sided Ore domains are good.

First recall that the following properties are equivalent for any domain R (cf. [FA, Ch. 9, Lemma 9.3.2, and the section between 7.16 and 7.17] and [ST, Ch. II]).

(a) R is right Ore, i.e. $sR \cap rR \neq 0$ for all r and s from R^o;

(b) there is a skew-field $Q \supseteq R$ - the so-called right skew-field of fractions of R - such that $Q = \{rs^{-1}: r,s \in R;\ s \neq 0\}$;

(c) there is a skew-field $Q \supseteq R$ such that for all $q_o,\ldots,q_{n-1} \in Q$ there is a $t \in R^o$ with $q_i t \in R$ for all $i < n$ (n any natural number).

It turns out that these conditions are equivalent to (0) $\rightarrow$ (-1), where the latter are the corresponding conditions from Lemma 7.4:

Lemma 1. A domain R is right Ore iff the conditions below are equivalent for all pp formulas ϕ of L_R.

(-1) $\phi(_RR) \neq 0$;

(0) $\phi(M) = M$ for all divisible left R-modules M.

Proof: Let first R be right Ore and Q its right skew-field of fractions. By Lemma 7.4 it suffices to show (0) $\rightarrow$ (-1). Let ϕ be the formula $\exists\bar{y} \bigwedge_{i<n} s_i x = \sum_{j<m} s_{ij} y_j$ and suppose (0) holds for ϕ. Being a skew-field, Q is divisible as a left R-module, too. Thus $\phi(_RQ) = Q$. Then $_RQ \models \phi(1)$, hence there are $q_j \in Q$ with $s_i = \sum_{j<m} s_{ij} q_j$ $(i < n)$. Applying (c) above choose $r_j \in R$ $(j < m)$ and $t \in R^o$ such that $s_i t = \sum_{j<m} s_{ij} r_j$ $(i < n)$.

Then $0 \neq t \in \phi(_RR)$, i.e. ϕ satisfies (-1), too.

Clearly, condition (a) above holds iff $\phi(_RR) \neq 0$ for all ϕ of the form $\exists y_o y_1\ (x = sy_o \wedge x = ry_1)$, where $r,s \in R^o$.

So, for the converse it is enough to show that $\phi(M) = M$ for all divisible left R-modules M and all ϕ as above. However, this is trivial. □

Lemma 2. If R is a right and left Ore domain, then condition (-1) is equivalent to the condition

(2) $\phi(N) = N$ for all injective left R-modules N.

Proof: As the first half of the above proof, noticing that Q, the right skew-field of fractions of R, is injective as a left R-module if R is also left Ore (cf. the literature cited above), for then (2) is enough to derive (-1). □

Consequently, for a right and left Ore domain conditions (-1) through (5) from Lemmata 7.2 and 7.4 are equivalent. In particular, these are good.

PART III. THE ZIEGLER SPECTRUM

Ivo Herzog

In this part of the paper we investigate the closed subset I(T) of the Ziegler Spectrum when the theory T has a regular generic.

§9. Pure-injectives

I mention a few basic facts about pure-injective modules and pure-injective hulls, all of which may be found in [PR] or [Zie].

Definition 1a. A partial map f from M to N for which

$$M \models \varphi(\bar{a}) \underset{(\Leftrightarrow)}{\Rightarrow} N \models \varphi(f\bar{a})$$ for all $\bar{a} \in \operatorname{dom} f$ and ppf $\varphi(\bar{x})$ is called

a **partial homomorphism (isomorphism)** from M to N.

b. An embedding f of M into N is called **pure** if it is a partial isomorphism.

Definition 2. A module M is **pure-injective** if every homomorphism f from a pure submodule N′ of N to M can be lifted to a homomorphism $\tilde{f}$ from N to M:

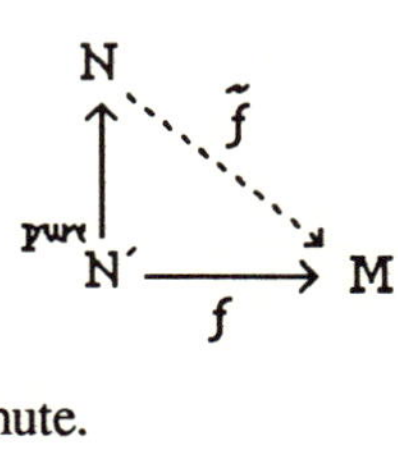

making the above diagram commute.

Definition 3. Let M be a pure-injective module and $M \supseteq A$. Then $\mathbf{H_M(A)}$ is a **pure-injective hull of A in M** if:

a) $H_M(A)$ is a pure-injective pure submodule of M containing A and
b) If B is a pure-injective pure submodule of M and $H_M(A) \supseteq B \supseteq A$, then $B = H_M(A)$.

The most basic properties of pure-injective modules are given by the following facts:

Fact 1. For M a pure-injective module and $M \supseteq A$, a (pure-injective) hull $H_M(A)$ of A in M exists.

Fact 2. a. M is pure-injective iff every partial homomorphism f from N to M lifts to a homomorphism $\tilde{f}$ from N to M.

b. If f is a partial isomorphism from N to M and N is pure-injective, then $\tilde{f}$ restricted to $H_N(\mathrm{dom}\, f)$ is a pure imbedding.

A pp-complete type p(x) is one of the form $tp^{\pm}(A) = tp^{+}(A) \cup tp^{-}(A)$. If the theory T (a complete extension of T_R) is understood, we will denote by $S_I^{\pm}(C)$ the set of T-consistent pp-complete types over C in the set of variables indexed by I. Note that by elimination of quantifiers $S_I^{\pm}(C)$ is naturally homeomorphic to $S_I(C)$.

From Facts 1 and 2 we see that for a pure-injective M and $M \supseteq A$, $H_M(A)$ is determined by $tp^{\pm}(A/\emptyset)$ up to isomorphism. Thus it makes sense to speak about H(p) when p(x) is a pp-complete type. It also allows us to sometimes forget the subscript in " $H_M(A)$."

From Fact 2.b we can derive

Fact 3. If $c \in H_M(A)$, then $tp^{+}(c/A) \vdash_M tp(c/A)$.

Example: Let $R = Z$, the group of integers. Then the pure-injective indecomposables are the Prüfer groups $Z(p^{\infty})$ for each prime p, the p-adic completion of Z for each prime p, every indecomposable finite (abelian) group and $\mathbb{Q}$, the rationals.

§ 10. The Topology

If U is an indecomposable pure-injective module then for every $a \varepsilon U$, $a \neq 0$, we know that $U = H(a) = H(tp^{\pm}(a))$. Thus there are at most $2^{|R|}$ pure-injective indecomposables up to isomorphism. Denote by T_R^* the "largest" complete theory of R-modules, i.e. $Th(\oplus U^{(\aleph_0)})$ where the U's range over all isomorphism types of indecomposable pure-injective modules over R. For more on this see [PR, §2.6]. In general, for every T, a complete extension of T_R, we let I(T) be the set of pure-injective indecomposables which occur as direct summands of models of T. $I(T_R^*)$ is just the set of all pure-injective indecomposable R-modules up to isomorphism.

In [Zie], Ziegler defines a topology on $I(T_R^*)$, a basis of which is the sets of the form $(\varphi/\psi) = \{ U \varepsilon I(T_R^*) : |\varphi(x)/\psi(x)| > 1 \}$ and the closed sets of which are exactly those of the form I(T), T a complete theory of modules. That this is indeed a topology will follow from the proposition below.

We work in T_R^*. Let $I = \{ p \varepsilon S_1^{\pm}(\emptyset) :$ p is indecomposable, i.e. H(p) is } (§0, Fact 2) endowed with the relative subspace topology which it inherits from the Stone space of all 1-types (in this space a clopen basis is given by $\{ [\sigma(x)] = \{p: \sigma(x) \varepsilon p\}: \sigma(x)$ a boolean combination of ppfs.}). From Fact 2 of §0, one can deduce the following

Lemma [Zie,Cor.4.5]. If Δ is a finite subset of an indecomposable pp-complete type p, then there are $\varphi/\psi \varepsilon p$, i.e. $\varphi, \neg\psi \varepsilon p$, such that

$T_R^* \vdash (\varphi \wedge \neg\psi) \rightarrow \bigwedge \Delta.$

In the following proposition we paraphrase [Zie, Thm. 4.6].

Proposition. The sets (φ/ψ) form a basis for the quotient topology on $I(T^*_R)$ induced by the map $H\colon I \to I(T^*_R)$ which takes an indecomposable pp-complete type p into the isomorphism type of its hull H(p).

Proof: First we show that (φ/ψ) is open, i.e. that $H^{-1}(\varphi/\psi)$ is. Let $U \varepsilon (\varphi/\psi)$ and take $p(x)$ with $H(p) \cong U$. We need to find an open subset O of I such that $p \varepsilon O$ and $(\varphi/\psi) \supset H(O)$. Let $a \models p(x)$ and $c \varepsilon H(p)$ so that $H(p) \models \varphi(c) \wedge \neg \psi(c)$. By Fact 3 of § 9 we can take $\sigma(x,y) \varepsilon tp^+(c,a)$ such that $\sigma(x,a) \vdash_{H(p)} \varphi(x) \wedge \neg\psi(x)$.

Thus $\dfrac{\exists x\,(\sigma(x,y) \wedge \varphi(x))}{\exists x\,(\sigma(x,y) \wedge \psi(x))} \varepsilon\ tp^{\pm}(a)$. But if $U \varepsilon \left(\dfrac{\exists x\,(\sigma(x,y) \wedge \varphi(x))}{\exists x\,(\sigma(x,y) \wedge \psi(x))}\right)$, then clearly $U \varepsilon (\varphi/\psi)$; so take

$O = [\, \exists x\,(\sigma(x,y) \wedge \varphi(x)) \wedge \neg\, (\exists x\,(\sigma(x,y) \wedge \psi(x)))\,]$.

On the other hand, let O be an open subset of $I(T^*_R)$ in the quotient topology induced by H. So $H^{-1}(O)$ is open and if $H(p) \varepsilon O$, there is a formula $\varphi(x) \wedge \bigwedge_{i<n} \neg\psi_i(x)$ such that $p \varepsilon [\, \varphi(x) \wedge \bigwedge_{i<n} \neg\psi_i(x)\,]$ and $H^{-1}(O) \supseteq [\, \varphi(x) \wedge \bigwedge_{i<n} \neg\psi_i(x)\,]$. By the Lemma there are $\sigma(x)$ and $\chi(x)$ such that $p \varepsilon [\, \sigma(x) \wedge \neg\chi(x)]$ and $[\, \varphi(x) \wedge \bigwedge_{i<n} \neg\psi_i(x)\,] \supseteq [\, \sigma(x) \wedge \neg\chi(x)]$. Therefore $H(p) \varepsilon H([\, \sigma(x) \wedge \neg\chi(x)]) = (\sigma/\chi)$ and $O \supseteq (\sigma/\chi)$. Thus we have shown that each open subset in the quotient topology is a union of sets of the form (φ/ψ). □

The closed set I(T) turns out to be a rather important invariant of T. One can certainly perform a Cantor-Bendixson analysis on it. Given any topological space X its CB derivative X′ is just $X \setminus \{$ isolated points of X $\}$ and the higher CB derivatives $X^{(\alpha)}$ are defined by recursion on the ordinals as follows:

i. $X^{(0)} = X$

ii. $X^{(\beta+1)} = (X^{(\beta)})'$ and

iii. $X^{(\lambda)} = \bigcap_{\alpha<\lambda} X^{(\alpha)}$ if λ is a limit ordinal.

CB(X), the CB rank of X, is then defined as the greatest α for which $X^{(\alpha)} \neq \emptyset$. If no such α exists then we say CB(X) = ∞. We also talk about rk(a), the CB rank of a point $a \in X^{(\alpha)}$ as the greatest α for which $a \in X^{(\alpha)}$.

Another measure of the complexity, which we call **m-dimension**, can be [cf. PR, Chap. 10] defined on pairs of ppfs φ and ψ when $\varphi \supseteq \psi$ as follows:

i. $\textbf{m-dim}_T\,(\varphi/\psi) \geq 0$ if $T \models \exists x\ (\varphi(x) \wedge \neg\, \psi(x))$

ii. $\textbf{m-dim}_T\,(\varphi/\psi) \geq \alpha + 1$ if there is a sequence $\{\varphi_n(x)\}_{n<\omega}$ such that $\varphi = \varphi_0 \supseteq \varphi_1 \supseteq \varphi_2 \supseteq \cdots \supseteq \varphi_n \supseteq \cdots \supseteq \psi$ and $\dim_T(\varphi_n/\varphi_{n+1}) \geq \alpha$ for all $n < \omega$ or if there is a sequence $\{\psi_n(x)\}_{n<\omega}$ such that $\varphi \supseteq \cdots \supseteq \psi_n \supseteq \cdots \supseteq \psi_2 \supseteq \psi_1 \supseteq \psi_0 = \psi$ and $\dim_T(\psi_{n+1}/\psi_n) \geq \alpha$ for all $n < \omega$.

iii. $\textbf{m-dim}_T\,(\varphi/\psi) \geq \lambda$ if $\dim_T(\varphi/\psi) \geq \alpha$ for all $\alpha < \lambda$ if λ is a limit ordinal.

Then m-dim$_T(\varphi/\psi) = \alpha$ if m-dim$_T(\varphi/\psi) \geq \alpha$ and m-dim$_T(\varphi/\psi) \not\geq \alpha + 1$. If m-dim $(\varphi/\psi) \geq \alpha$ for all ordinals α then m-dim$_T(\varphi/\psi) = \infty$. By the dimension of a module M we mean m-dim$_{Th(M)}$ ($x = x$ / $x = 0$) if this is not ∞; otherwise we say M does not have m-dimension.

m-dimension and CB rank are then related by the next

Theorem [Zie, Thm.8.6]. If R is a countable ring then m-dim$_T(\varphi/\psi) = \max\{\, rk_{I(T)}(U) : U \in I(T) \cap (\varphi/\psi) \,\}$ and hence $rk_{I(T)}(U) = \min\{$ m-dim$_T(\varphi/\psi) : U \in (\varphi/\psi)\, \}$.

§ 11. The Hull of the Generic

In this section we consider a theory T with a regular generic 1. It is shown that if R is countable and T has m-dimension, i.e. that $CB(I(T)) < \infty$, then H(1) is the unique unlimited point in I(T) of maximal CB rank.

Suppose that R is a commutative noetherian domain. There is a bijection between Spec(R) and the set of indecomposable injective R-modules given by $\mathfrak{p} \rightarrow E(R/\mathfrak{p})$ - E denotes here the injective hull of an R-module. Let $T_{inj} = Th(\oplus E(R/\mathfrak{p})^{(\aleph_0)} : \mathfrak{p} \in Spec(R))$, the largest theory of injective R-modules. By [E-S], injectivity is an elementary property so every model of T_{inj} is injective as are all of its indecomposable direct summands. $I(T_{inj})$ is thus exactly the image of Spec(R) in $I(T^*_R)$ under the above imbedding. Prest has noted [PR,§4.7, Example 2 and §6.I] that this imbedding takes closed subsets of Spec(R) into open subsets of $I(T_{inj})$. Specifically if $\mathfrak{p}$ is a prime ideal then the image of $V(\mathfrak{p}) = \{ \mathfrak{q} \in Spec(R) : \mathfrak{q} \supseteq \mathfrak{p}\}$ is the open (in $I(T_{inj})$) subset $(\mathfrak{p}x = 0 / x = 0)$. So while the maximal ideal $\mathfrak{m}$ is a closed point in Spec(R) ($\{\mathfrak{m}\} = V(\mathfrak{m})$), in $I(T_{inj})$ it is isolated by the neighborhood $(\mathfrak{m}x = 0 / x = 0)$.

Since R is a domain, Spec (R) is an irreducible space whose generic point is the prime ideal 0. The point in $I(T_{inj})$ to which 0 corresponds is E(R), the field of quotients of R. E(R), being a field, is a pp-simple indecomposable injective (and hence pure-injective) module. So any two non-zero elements of E(R) have the same type 1 and E(R) = H(1).

Let $c \in E(R/\mathfrak{q})$. Since $E(R/\mathfrak{q})$ is indecomposable and $E(R/\mathfrak{q}) \supseteq Rc$ it follows that $E(R/\mathfrak{q}) \cong E(Rc)$. But the homomorphism $f: R \rightarrow Rc$ for which $f(1) = c$ lifts to a homorphism $\tilde{f}: E(R) \rightarrow E(Rc) \cong E(R/\mathfrak{q})$. The existence of such a map implies that $tp^+(c) \supseteq 1^+$ and that indeed for every $a \in M \models T_{inj}$, $tp^+(a) \supseteq 1^+$.

The pp-type 1^+ carries in it the least amount of positive information. In fact, $\varphi(x) \in 1^+$ iff $\varphi(M) = M$ for every $M \models T_{inj}$. Also note that each indecomposable in $I(T_{inj})$ is unlimited in T_{inj} so that T_{inj} has no pp-definable subgroups of finite index. p is thus the generic type. By Fact 1 of §2, $\mathbb{1}$ is regular and we have naturally associated the generic point of Spec(R) with the hull of generic type (a regular type, in this case) of T_{inj}.

Theorem. Suppose that R is countable and T has a regular generic p. Then $H(\mathbb{1})$ is the unique unlimited $U \in I(T)$ which is a closed point. In particular, if $0 < CB(I(T)) = \alpha < \infty$, then $H(\mathbb{1})$ is the only $U \in I(T)$ for which $rk(U) = \alpha$.

Proof: Let $U = H(\mathbb{1})$. By Lemma 2(3) of §5, U is pp-simple so $End_R U \cong \Delta$, a division ring and U is a one-dimensional (right) vector space over Δ. Now it is well known that $End_\Delta U \cong \Delta$, if we let it act on the left. From now on, we shall think of Δ as acting on the left; if we want to about the action of Δ on the right, we will denote Δ by Δ_r. We also see that $\bar{R} = R/fin_R T$ imbeds into Δ for every element $r \in R$ commutes with the action of $End_R U \cong \Delta_r$ and if $s \in fin_R T$ then $sU = 0$.

Let D be the division ring in Δ generated by $\bar{R}$. Let $\delta_1,\ldots,\delta_\alpha$ be a basis for Δ over D i.e. $\Delta = \bigoplus_{i<\alpha} D\delta_i$ as a (left) vector space over D. Then for $a \in U$, $U = \Delta a = \bigoplus_{i<\alpha} D\delta_i a$ as an $\bar{R}$-module since $D \supseteq \bar{R}$. But U is indecomposable so $\alpha = 1$ and $\Delta = D$, the division ring (in Δ) generated by $\bar{R}$.

Let $V \in I(T)$ be unlimited. Then if $\varphi(x) \in \mathbb{1}^+$, it means that $\varphi(V) = V$. Suppose moreover that $I(Th(V)) = \{V\}$ i.e. that V is a closed point in I(T). By the Theorem in §10, V has dimension zero. V has neither an ascending nor descending chain of pp-definable subgroups so we get a composition series of pp-definable subgroups of V. Using a Jordan-Hölder argument one can show that the length of a composition series is an invariant of V which we call $\mu(V)$, the multiplicity of V.

Claim: For each $\delta \varepsilon \Delta$ there is a ppf. $\sigma_\delta(x,y)$ such that $y = \delta x$ iff $U \models \sigma_\delta(x,y)$ and $\sigma_\delta(x,y)$ defines in V the graph of a bijective $\mathbb{Z}$-homomorphism (also denoted by δ).

Proof: We prove the claim for all $r \varepsilon \bar{R}$ and then show that the set of δ for which the claim holds is a division ring.

i. If $r \varepsilon \mathrm{fin}_R T$ then $rx = 0 \ \varepsilon \ 1^+$ and so $V \models rx = 0$ making V an $\bar{R}$-module. If $r \varepsilon \bar{R}$, $y = rx$ is a ppf. defining the graph of r. If $r \neq 0$ then $rU = U$ since U is pp-simple and so $r \mid x \ \varepsilon \ 1^+$. But then $V \models r \mid x$ and $rV = V$. Since $rV \cong V/\ker r$, $\mu(V/\ker r) = \mu(rV) = \mu(V)$ forcing $\ker r = 0$.

ii. If $\delta_1, \delta_2 \ \varepsilon \ \Delta$ satisfy the claim then:

$\sigma_{\delta_1+\delta_2}(x,y)$ is defined by the ppf. $\exists z \, \sigma_{\delta_1}(x,z) \wedge \sigma_{\delta_2}(x,y-z)$

$\sigma_{\delta_1\delta_2}(x,y)$ is defined by the ppf. $\exists z \ \sigma_{\delta_2}(x,z) \wedge \sigma_{\delta_1}(z,y)$ and

$\sigma_{\delta_1^{-1}}(x,y)$ is defined by $\sigma_{\delta_1}(y,x)$.

So if δ is $\delta_1 + \delta_2$, $\delta_1\delta_2$ or δ_1^{-1} then $\sigma_\delta(x,y)$ is a ppf. such that $y = \delta x$ (in U) iff $U \models \sigma_\delta(x,y)$ and by hypothesis it clearly defines a function on V. If $\delta \neq 0$, then $U \models \exists x \, \sigma_\delta(x,y)$ so $\exists x \, \sigma_\delta(x,y) \ \varepsilon \ 1^+$ and $V \models \exists x \, \sigma_\delta(x,y)$ i.e. $\delta V = V$. As in case i., $\ker \delta = 0$, δ is bijective and the claim is verified.

We need to show that V is a left vector space over Δ. Let $\alpha(x)$ be the ppf. $\exists y \exists z \, (\sigma_{\delta_1+\delta_2}(x,y) \wedge \sigma_{\delta_1}(x,z) \wedge \sigma_{\delta_2}(x,y-z))$ and let $\mu(x)$ be the ppf. $\exists y \exists z \, (\sigma_{\delta_1\delta_2}(x,y) \wedge \sigma_{\delta_2}(x,z) \wedge \sigma_{\delta_1}(z,y))$. Evidently, $V \models \alpha(x)$ means that for all x in V, $(\delta_1 + \delta_2)(x) = \delta_1 x + \delta_2 x$ and $V \models \mu(x)$ means that for all $x \ \varepsilon \ V$ $(\delta_1\delta_2)x = \delta_1(\delta_2 x)$. But $U \models \alpha(x) \wedge \mu(x)$ so $\alpha(x) \wedge \mu(x) \ \varepsilon \ 1^+$ and therefore $V \models \alpha(x) \wedge \mu(x)$ and V is now a vector space over Δ. Since $\Delta \supseteq \bar{R}$ and V is indecomposable, V is one-dimensional over Δ.

Now let $a \ \varepsilon \ U$, $b \ \varepsilon \ V$, $a,b \neq 0$. $tp^+(b) \supseteq 1^+ = tp^+(a)$ so there is a partial map f from U to V such that $f(a) = b$. As V is pure-injective this lifts to a map $\tilde{f}:U \to V$ which commutes with the left action of Δ so it must be an isomorphism and $U \cong V$.

U is closed in I(T) since its closure is just I(Th(U)) = {U}. By the above it follows that if V, unlimited, is another such closed point of I(T) then $U \cong V$. If $0 < CB(I(T)) = \alpha$, $I(T)^{(\alpha)}$ is a closed discrete set so all of its points are closed and unlimited. But that means that $I(T)^{(\alpha)} = \{U\}$ and the theorem is proved. □

The above theorem does not generalize to abelian structures. For example, if we consider the theory T of a torsion-free divisible abelian group and we include in the language a unary predicate P which interprets in a model of the same theory a non-trivial divisible subgroup, then I(T) will consist of two indecomposable pure-injective abelian structures: a copy of the rationals in which P interprets the trivial group 0 and a copy of the rationals again with P interpreting the whole group. It is the former of the two which is the hull of the generic so we see that the generic is regular. But both points of I(T) are T-unlimited and closed.

References.

[B-R] Baudisch, A. and Rothmaler, Ph., The stratified order in modules, pp. 44-56 in Frege Conference, Schwerin 1984, Math. Res, Akademie-Verlag, Berlin, 1984.

[E-S] Eklof, P. and Sabbagh, G., Model-completions and modules, Ann. Math. Logic, 2(3) (1971), 251-295.

[FA] Faith, C., Algebra: Rings, Modules and Categories, Springer-Verlag, Berlin New-York, 1973.

[Ga] Garavaglia, S., Forking in modules, Notre Dame J. Formal Logic, 22 (1981), 155-162.

[Po] Poizat, B., Groupes stables, avec types génériques réguliers, J. Symbolic Logic, 48(2) (1983), 339-355.

[P-P 1] Pillay, A. and Prest, M., Forking and pushouts in modules, Proc. London Math. Soc. Ser. 3., 46 (1983), 365-384.

[P-P 2] Pillay, A. and Prest, M., Modules and stability theory, Trans. Amer. Math. Soc., 300(2) (1987), 641-662.

[Pr] Prest, M., The generalized RK-order, orthogonality and regular types for modules, J. Symbolic Logic, 50(1) (1985), 202-219.

[PR] Prest, M., Model Theory and Modules, London Math. Soc. Lecture Note Series. 130, Cambridge Univ. Press, 1988.

[Ro 1] Rothmaler, Ph., Stationary types in modules, Z. math. Logik Grundl. Math., 29 (1983), 445-464.

[Ro 2] Rothmaler, Ph., On a question of Bruno Poizat, J. Symbolic Logic, 52(1) (1987), 340-341.

[Ro 3] Rothmaler, Ph., On a certain ideal, pp. 309-312 in Contributions to General Algebra, Hölder-Pichler-Tempsky (Vienna) and Teubner (Stuttgart), 1987.

[Ro 4] Rothmaler, Ph., Review of [Zie] in Zentralblatt Math. 593.16019.

[SH] Shelah, S., Classification Theory and the Number of Non-somorphic Models, North-Holland, Amsterdam, 1978.

[ST] Stenström, B., Rings of Quotients, Springer-Verlag, Berlin New-York 1975.

[Zie] Ziegler, M., Model theory of modules, Ann. Pure Appl. Logic, 26(2) (1984), 149-213.

MODEL THEORETIC VERSIONS OF WEIL'S THEOREM ON PREGROUPS

Elisabeth Bouscaren

In 1955, A. Weil published a paper ["On algebraic groups of transformations", Am. J. of Math., vol.77 (1955), p:355–391] where, starting from a variety V over some algebraically closed field K, together with a binary operation on V which has "good" properties (associativity, rationality) on a large piece of V (generic points), he constructs an algebraic group G over K, whose multiplication is an extension of the given one on generic points and which is birationally equivalent to V.

More precisely:

Let K be an algebraically closed field and let V be an irreducible variety over K such that there is a mapping f: VxV $\rightarrow$ V with the following properties:

(i) if a,b are independent generic points of V over K, then

$$K(a,b) = K(a,c) = K(b,c)$$

(ii) if a,b,c are independent generic points of V over K, then

$$f(f(a,b),c) = f(a,f(b,c)).$$

Then there is an algebraic group G over K which is birationally equivalent to V, such that this birational equivalence takes f(a,b), for a,b independent generics of V, to the product of the images of a and b.

Model–theorists working on stable groups became interested in this theorem in the following context: first, recall that, by a stable (ω–stable) group, we mean a group $(G,\cdot)$ definable in M^n for M a model of a stable (ω–stable) theory or interpretable, i.e. definable on some quotient of M^n by some definable equivalence relation.

Amongst the first natural examples of ω–stable groups are algebraic groups over an algebraically closed field K (they are definable in the theory of K in the language of fields).

Some years ago arose the conjecture that in fact all simple ω–stable groups with finite Morley rank "were" algebraic groups. Now, if one hopes to be able to define a topology and a variety structure on any such abstract group, one should certainly first try to do it (and the construction should hopefully be rather canonical) in the particular case of a group interpretable in some algebraically closed field but which has a priori no variety structure which makes it into an algebraic group.

This question was asked by B. Poizat and a first positive answer was given at the time by L. van den Dries (unpublished notes, characteristic 0 case): in order to simplify, let us say that the idea is to find a good $V \subseteq G$ with a variety structure satisfying the assumptions of Weil's theorem and then, to get the the algebraic group by applying the theorem.

This was unsatisfactory, even in the characteristic 0 case, for two reasons: first, if one does not know the proof of Weil's theorem, then one does not really know much about this structure of algebraic group and the way it relates to the original group; secondly, using the fact that we start with an actual group, there should be a more direct proof, avoiding some of the difficulties encountered when starting with an operation defined only on generic points.

This indeed turned out to be the case: a direct proof was given by E. Hrushovski (1986), in all characteristics.

This is the proof we want to present here.

Theorem 1:

Let K be an algebraically closed field, let $(G,\cdot)$ be interpretable in K, then G is definably isomorphic to an algebraic group over K.

More precisely this is decomposed in two parts:

Theorem 1–A: Let $(G,\cdot, \mathrm{inv})$ (inv denotes the inverse on G) $\subseteq K^n$ be a group definable with parameters in some countable $k_0 < K$, such that for a,b generic independent,

$$a.b \in k_0(a,b)$$

$$\mathrm{inv}(a) \in k_0(a).$$

Then there is definably in K a structure of variety on G (over K) which makes $(G,\cdot,\mathrm{inv})$ into an algebraic group.

Theorem 1–B: Let H be interpretable in K, then there is $G \subseteq K^n$ satisfying the assumptions of 1–A, such that H and G are definably isomorphic.

The theorem above certainly qualifies as a model–theoretic version of Weil's theorem, but it does not deal with the part of the theorem which constructs a group from an operation on the generic points. Now, the following result can certainly be considered as the model–theoretic version of this aspect of Weil's theorem. It was in fact proved by E. Hrushovski prior to the other one, and is purely model–theoretic.

Theorem 2: (Hrushovski, Ph.D., Berkeley, 1986):

Let T be an ω–stable theory, let $p \in S(\emptyset)$ be a stationary type and let $*$ be a partial $\emptyset$ – definable operation such that:

(i) for a,b realizing p, independent,

$a*b$ realizes $p_{|a}$ and $p_{|b}$ (where $p_{|a}$ denotes the unique non forking extension of p over a)

(ii) for a,b,c, realizing p, independent,

$(a*b)*c = a*(b*c)$.

Then there is a definable set G, a definable operation $\cdot$ on G and a definable embedding g of p into G, such that

$(G,\cdot)$ is a group

for $a,b \models p$, independent, $g(a*b) = g(a) \,.\, g(b)$

g(p) is the generic of G.

(In fact this theorem with the weaker conclusion that G be infinitely definable is proved for all stable theories).

We will not say more about this aspect, but one should note that, from these two theorems, one recovers the full statement of Weil's theorem: let V be an irreducible variety satisfying the assumptions in Weil's theorem and consider p the generic type of V. Then p satisfies the assumptions in Theorem 2, and by applying first Theorem 2 and then Theorem 1, one gets the algebraic group.

Remark: The same kind of result was also more recently proved in a different (and unstable) setting by A. Pillay ["On groups and fields definable in

O–minimal structures", preprint]. In particular he shows that if a group G is definable in the reals, then G is a Lie group.

Proof of Theorem 1:

We are going to assume that the group G is connected but this is no loss of generality as the general case follows from the connected case.

Theorem 1–A:

Let (G,·, inv) $\subseteq K^n$ be a connected definable group with parameters in $k_o < K$such that, for a,b generic independent

$$a.b \in k_o(a,b)$$
$$inv(a) \in k_o(a),$$

then there is definably in K a structure of variety on G which makes (G,·,inv) into an algebraic group.

We are going to need the following easy lemma:

Lemma 0:

a) – Let V be an irreducible variety and let $X \subseteq V$ be a definable set, X containing the generic of V. Then X contains an open subset O of V (and of course O contains the generic).

b) – Let V and V' be two irreducible varieties, and let f be a definable map from V to V' such that, on the generic of V, f is rational. Then there is $O \subseteq V$, open, such that f|O is a morphism.

Proof:

a) – The set X is definable, therefore it is a finite union of sets of the form $O \cap F$, where O is open in V and F is closed in V. Choose one such $O \cap F$ that contains the generic, so F contains the generic, but the complement of F in V is open and, as V is irreducible, must also contain the generic, so F = V.

b) – Choose a definable $X \subseteq V$, containg the generic, such that f(a) is a given rational function of a, for all a in X. By a) X contains an open set of V.

Proof of Theorem 1–A:

The group G is definable, so $G = \bigcup_{i<m} (O_i \cap F_i)$, where we can assume the O_i's to be principal open sets in K^n and where the F_i's are closed in K^n. Let V_o be one of these intersections, containing the generic of G, p. Then on V_o we have the structure of an irreducible prevariety, with generic p; we also have the usual structure of prevariety on $V_o \times V_o$, with generic pxp. Let X be a definable subset of $V_o \times V_o$ containing pxp, such that if $(x,y) \in X$, then x.y is rational over x,y. Let $X' = \{(x,y) \in X ; x.y \in V_o\}$, X' is definable. By the lemma above, there is $M_o \subseteq X'$, open in $V_o \times V_o$, such that multiplication, from M_o in V_o, is a morphism. For the same reasons, there is $V_1 \subseteq V_o$, open such that inv is a morphism from V_1 in V_o.

Now let

$Y = \{x \in V_1$; for all y generic independent from x, $(y,x) \in M_o$ and $(inv(y),y.x) \in M_o\}$.

By definability of the type p, Y is a definable set, and Y contains the generic p. By the lemma again, there is $V_2 \subseteq V \subseteq V_1$, open, and of course, inv is still a morphism from V_2 in V_o. Now let $V = V_2 \cap inv(V_2)$, then V is open, because V is the inverse image (in V_2) by a morphism, of an open set, and $V = inv(V)$. Let $M = \{(x,y) \in M_o \cap VxV ; x.y \in V\}$, again, because multiplication is a morphism, M is open.

So, by taking smaller and smaller open sets we have come to the following situation: we have V, open in V_o, therefore with the induced variety structure, and M, open in VxV, such that:

(i) multiplication is a morphism from M into V

(ii) inv is a morphism from V into V and $inv(V) = V$

(iii) for all x in V, for all y generic independent from x, (y,x) and (inv(y),y.x) are both in M.

The structure of variety on G is obtained by covering G by translates of V (i.e. of the form a.V). As G is an ω–stable group and V contains the generic of G, we know that a finite number of translates of V will be sufficient to of G, we know that a finite number of translates of V will be sufficient to cover G.

In order to see that this indeed gives G the structure of a variety and in fact of an algebraic group, we need the following lemma:

Lemma:

Let $a,b \in G$, let $H = \{(x,y) \in VxV;\ a.x.b.y \in V\}$. Then

– H is open

– the map f_{ab} from H into V, which takes (x,y) to a.x.b.y is a morphism.

Proof of the Lemma:

Let $(x_o,y_o) \in H$, we want to find H_o, $(x_o,y_o) \in H_o \subseteq H$, open, such that f_{ab} restricted to H_o is a morphism.

We know that b = c.d, where c and d both realize the generic p; let e also realize p, independent from $\{a,c,d,x_o,y_o\}$. Let $H_o = \{(x,y) \in VxV;$ $(e.a,x) \in M$, $(e.a.x,c) \in M$, $(e.a.x.c,d) \in M$ $(e.a.x.c.d,y) \in M$, $(inv(e),e.a.x.c.d.y) \in M\}$.

First, by the choice of e, and by applying condition (iii) on V each time, $(x_o,y_o) \in H_o$.

We see that H_o is open in VxV by applying successively the following classical facts: if O is open in VxV, if h is a morphism form O in V, if $z \in V$, then the set

$$\{(x,y);\ (x,z) \in O \text{ and } (h(x,z),y) \in O\}$$

is open, and also,

$$O_z = \{x \in V;\ (z,x) \in O\}$$

is open and h_z, from O_z in V, is a morphism.

Now a.x.b.y = inv(e).e.a.x.c.d.y , so $H_o \subseteq H$, and over H_o, f_{ab} becomes a composition of morphisms because at each step the elements one wants to multiply are in M, and hence it is a morphism.

We can now go back to the proof of the theorem. Choose $a_1,...,a_n$ in G such that $G = a_1 V \cup ... \cup a_n V$ (where aV denotes the set $\{a.x; x \in V\}$). In order to check that this, together with the left translations f_i from V into a_i V, is a prevariety on G, we need that for all i,j

– $V_{ij} = \{x \in V; a_i.x \in a_j V\} = \{x \in V; inv(a_j).a_i.x \in V\}$ is open

– the map f_{ij} from V_{ij} into V which takes x to $inv(a_j).a_i.x$ is a morphism.

But, it is a direct consequence of the lemma that, for all a in G, the set $V_a = \{x \in V; a.x \in V\}$ is open and the left translation by a is a morphism.

It remains to check that multiplication and inverse are morphisms.

Multiplication:

GxG, as a variety is covered by products of the form aVxbV, which get their variety structure from VxV. To say that multiplication is a morphism means exactly that the set

$A_{abc} = \{(x,y) \in VxV; a.x.b.y \in cV\} = \{(x,y) \in VxV; inv(c).a.x.b.y \in V\}$

is open in VxV and that the map from A into V which takes (x,y) to inv(c).a.x.b.y is a morphism. This is exactly the lemma.

inverse:

It is a morphism if the set

$A_{ab} = \{x \in V; inv(a.x) \in bV\} = \{x \in V; inv(a.x.b) \in V\}$

is open and the map from A_{ab} into V which takes x to inv(a.x.b) is a morphism. But (condition (ii)) inv(V) = V, so $A_{ab} = \{x \in V; a.x.b \in V\}$, and again it is open, as a direct consequence of the lemma, and the map taking x to a.x.b is a morphism. We also have that, on V, inv is a morphism, so inv(a.x.b) is the composition of two morphisms. □

Theorem 1–B:

Let (H,·,inv) be a connected group interpretable in K. Then there is a definable group $(G,*, inv') \subseteq K^n$ and some countable $k_o \leq K$, k_o containing the defining parameters of H and G, such that H and G are definably isomorphic and, for a,b generic independent in G, $a*b \in k_o(a,b)$ and inv' (a) $\in k_o(a)$.

Proof:

Note first that, by elimination of imaginaries in algebraically closed fields, any interpretable group is definably isomorphic to some definable group in some K^n, so we can assume that $(H,\cdot, inv) \subseteq K^n$. Without loss of generality, assume K is very staurated.

Now if K has characteristic 0, then there is nothing left to prove, as any definable function is locally rational, so we assume that K has characteristic p>0.

Let $k \subseteq K$ be an uncountable algebraically closed field, containg all the defining parameters of H. There is some $q = 1/p^m$ such that, for all $\bar{a}, \bar{b} \in H$, $\bar{a}.\bar{b} \in k(\bar{a}^q, \bar{b}^q)^n$ and $inv(\bar{a}) \in k(\bar{a}^q)^n$ (where if $\bar{a} = (a_1,...,a_n)$, $k(\bar{a}^q)$ denotes $k(a_1^q,...,a_n^q)$).

Let $\bar{a}$ realize the generic of H over k. We define:

$k^*(\bar{a}) = k(\bar{a}^q, inv(\bar{a})^q, \bar{b}_1.\bar{a}.\bar{b}_2, \bar{b}_1.inv(\bar{a}).\bar{b}_2; \bar{b}_1, \bar{b}_2 \in H \cap k^n)$.

We have that $k^*(\bar{a}) \subseteq k(\bar{a}^{q'})$, with $q' = q^2$.

Now $k(\bar{a}^{q'})$ is a finite extension of $k(\bar{a})$ hence so is $k^*(\bar{a})$, so there are $c_1,...,c_k$ in $k(\bar{a}^{q'})$, such that $k^*(\bar{a}) = k(\bar{a},c_1,...,c_k)$, and of course, each c_i is definable over $k \cup \bar{a}$.

Consider f: $H \rightarrow k^{n+k}$, definable injection such that for $\bar{a}$ generic,

$f(\bar{a}) = (\bar{a}, c_1, \ldots, c_k)$.

Trivially, $k^*(\bar{a}) = k^*(\text{inv}(\bar{a}))$, so $k(f(\bar{a})) = k(f(\text{inv}(\bar{a})))$, so if G is the image of H by f, with the obvious group law, it is true that, on a generic of G, the inverse is rational.

Now it is also trivial that, if $\bar{b} \in k^n \cap H$, then

$$k^*(\bar{a}.\bar{b}) = k^*(\bar{a}), \text{ so } k(f(\bar{a}.\bar{b})) = k(f(\bar{a}))$$

$$(*)$$

$$k^*(\bar{b}.\bar{a}) = k^*(\bar{a}), \text{ so } k(f(\bar{b}.\bar{a})) = k(f(\bar{a})).$$

We also have that, as f is a definable bijection, for $\bar{a}$, $\bar{b}$ generic, $f(\bar{a}.\bar{b}) \in k(f(\bar{a})^r, f(\bar{b})^r)^{n+k}$ for $r = 1/p^{\ell}$, for some ℓ.
Let k_o, countable, $k_o < k$, contain all the necessary parameters.
Let $\bar{b} \in H \cap k^n$ realize the generic of H over k_o, and let $\bar{a}$ realize the generic of H over k. By $(*)$, $f(\bar{a}.\bar{b}) \in k(f(\bar{a}))^{n+k}$, and we also have that

$$f(\bar{a}.\bar{b}) \in k_o(f(\bar{a})^r, f(\bar{b})^r)^{n+k}.$$

But, $k(f(\bar{a})) \cap k_o(f(\bar{a})^r, f(\bar{b})^r) = k_o(f(\bar{a}), f(\bar{b})^r)$: because $f(\bar{a})^r$ remains over $k(f(\bar{a}))$ of the same degree as over $k_o(f(\bar{a}), f(\bar{b})^r)$, since $\bar{a}$ is independent from k, which contains $f(\bar{b})^r$), over k_o.

Symmetrically, because $\bar{a}\wedge\bar{b}$ and $\bar{b}\wedge\bar{a}$ have the same type over k_o, we have that $f(\bar{a}.\bar{b}) \in (k_o(f(\bar{a}), f(\bar{b})^r) \cap k_o(f(\bar{a})^r, f(\bar{b})))^{n+k}$.

But these two fields are linearly disjoint over $k_o(f(\bar{a}), f(\bar{b}))$: more generally, it is classical algebra that if K_1, K_2 are linearly disjoint over k_o, if $x \in K_1$, $y \in K_2$, then $K_1(y)$ and $K_2(x)$ are linearly disjoint over $k_o(x,y)$. As $\bar{a}$ and $\bar{b}$ are independent over k_o, $k_o(f(\bar{a})^r) = K_1$ and $k_o(f(\bar{b})^r) = K_2$ are linearly disjoint over k_o, then we get the result by letting $f(\bar{a}) = x$ and $f(\bar{b}) = y$. It follows that $f(\bar{a}.\bar{b}) \in k_o(f(\bar{a}), f(\bar{b}))^{n+k}$, that is, that in G, the multiplication of two independent generics is rational. □

UA 753 – CNRS
Université Paris 7.

ON SUPERSTABLE FIELDS WITH AUTOMORPHISMS

Ehud Hrushovski

The Lie-Kolchin theorem states, essentially, that every connected solvable algebraic group over an algebraically closed field has a nilpotent derived group. This was generalized by Zil'ber and Nesin (independently) to groups of finite Morley rank. It was known that all ingredients of Nesin's proof generalize easily to superstable groups (satisfying an appropriate connectedness condition), except for the non-existence of definable groups of automorphisms of the field. The purpose of this note is to prove this fact: if F is a field, G a group of automorphisms of F, and (F,G) is superstable, then G = (1).

All groups are taken to be ∞-definable in $\mathbb{C}$, the universal domain of a superstable theory. We will use the notation of [M], and the theory of local weight in groups from [H, §3.3]. The basic definition is that of a regular type. The idea is that the elements of the group are co-ordinatized by n-tuples of realizations of the regular type. Thus for example if $A = (Z/2Z)^{\omega}$, with generic type p, then $B = A \times A$ is p-simple: an element of B is a pair of elements of A. But if $B = (Z/4Z)^{\omega}$, and A is identified with 2B, then B is not p-simple: an element of B can be analyzed in terms of p, but not in one step. Call a group p-connected if it is p-simple, connected, and has a generic type domination-equivalent to a power of p. One has the following existence property.

Fact 1. Let G be a group, H a group acting on G, and suppose the generic type of G is non-orthogonal to the regular type p. Then G has a normal, H-invariant subgroup N such that G/N is p-simple.

If p is chosen to have minimal possible U-rank, so that every forking extension of p is orthogonal to G, then every generic type of G/N will necessarily be $\equiv p^n$ for some n. In particular, every (superstable) group G has a filtration $G = G_n \supset G_{n-1} \supset \ldots \supset G_j = (1)$ such that each G_j is normal in G, and G/G_n is p-connected for some regular type p. So every simple group is p-connected for some p. The same is true for fields: If $G = G_a$ is the additive group of a field, then the multiplicative group $H = G_m$ acts transitively on G_a-(0). Since the filtration can be chosen to consist of H-invariant groups, it follows that n = 1 and G_a is p-connected.

Lemma 2 (Zil'ber): Let G be a p-connected group acting on the p-connected Abelian group V. Assume that V has no nontrivial G-invariant p-connected subgroups of smaller p-weight. Let $F = End_G(V)$. Then F is a definable field, and V is definably an F-vector space.

Proof: Let $a \in V$ - (0). By the indecomposability lemma (to be proved below), the subset of V generated by $\{x \cdot a : x$ realizes the generic type of G over $a\}$ by the operation $(u,v,w) \longrightarrow u - v + w$ is a coset of some ∞-definable subgroup of V. By minimality, this subgroup must be all of V. A fortiori, Ga generates V. Thus V is a simple Z[G]-module. By Schur's lemma, F is a division ring. Now F-(0) is the set of G-automorphisms of V: by [H, §4.2, lemma 2 and §3.4, theorem 2], it is definable. Hence so is F.

The use of the "indecomposability lemma" could have been avoided, but this seems pointless.

Proposition 3: Let F be a field, G a group of automorphisms of F, and assume that the structure (F,+,·,G, action of G on F) is superstable. Then G = 1.

Proof : If G is finite, then F must be algebraic over the fixed field F_0 of G. F_0 is a definable subfield, so it is algebraically closed; so $F = F_0$, i.e. $G = 1$.

Suppose G is infinite. Let F_0 be the algebraic closure of the prime field of F. Note that each element of F_0 has finite orbit under the action of G, so the connected component $G°$ of G must fix F_0 pointwise. Choose any element $\sigma \in G°$, $\sigma \neq 1$. We will show that the structure $(F,+,\cdot,\sigma)$ is already unsuperstable.

Case 1 F has characteristic 2.
Let $h(x) = \sigma(x)+x$. h is an additive endomorphism of F. The kernel K of h is precisely the fixed field of F. Let p be a regular type such that K is p-connected. As K is a proper subfield of F, we have $w_p(K) = 0$. By additivity of weight, $w_p(\text{range}(h)) = w_p(F)$; so by semi-regularity, the range of h is all of F. In particular, there exists $x \in F$ such that $h(x) = 1$. So $\sigma(x) = x + 1 \neq x$, but $\sigma^2(x) = \sigma(x+1) = x+2 = x$. Let $K_0 = \{x: \sigma x = x\}$, $K_1 = \{x: \sigma^2 x = x\}$. Then K_1 is an extension of degree 2 of K_0, contradicting the fact that K_0 is algebraically closed.

Case 2 F has characteristic other than 2.
Define h by: $h(x) = \sigma(x)/x$ for $x \neq 0$. Then h is a multiplicative endomorphism whose kernel is a proper subfield (minus $\{0\}$), so h is onto. Choose x such that $h(x) = -1$, and continue as in the previous case.

Problem 3. If σ is an automorphism of a field F, and (F,σ) is superstable, must σ be a power of the Frobenius automorphism? (Note that σ must have a finite fixed field by the above proof, so in particular F has prime characteristic).

Proposition 4: Let G be a p-connected group acting on the p-connected Abelian group V. Let N be a normal Abelian subgroup of G. Assume that V,G,N are non-trivial, and that every proper G-invariant p-connected subgroup of V is trivial. Let $R = \text{End}_N(V)$ and let F be the center of R.

Then F is a definable field, V is (definably) a finite-dimensional F-vector space, and G acts F linearly on V.

Proof: Let U be a p-connected, N-invariant subgroup of V of least possible non-zero p-weight. By the minimality of A and the finiteness of p-weight, V is a finite direct sum of G-conjugates of U. It follows that R is isomorphic to the n×n matrix ring over $End_N(U)$, so the center F of R is isomorphic to $End_N(U)$. By lemma 2, F is a definable field; so R is definable. Since N is Abelian, each $\sigma \in N$ acts on V as an N-endomorphism. This gives an embedding of N into the center of R, i.e. into F. In particular, $w_p(F) > 0$, so V is finite-dimensional as an F-space. R and its action on V are clearly 0-definable, hence so is F=center(R). Thus G acts on F naturally. By proposition (3), the action is trivial. So G acts F-linearly on V.

The rest of Nesin's proof of the Lie-Kolchin theorem is routine.

We now present a version of Zil'ber's indecomposability theorem in the context of regular types; Lascar and Berline have proved it at the same level of generality using U-rank, so it is only presented in order to demonstrate the technique. A subset C of G is a (left) coset of some subgroup if and only if C is closed under the operation: $(x,y,z) \rightarrow xy^{-1}z$. If this is the case, then C is a right translate of a unique subgroup S of G, namely $S = CC^{-1}$. If C is (∞-) definable, then so is S, and a type of C is called generic iff it is a translate of a generic type of S. The coset generated by a subset X of G is by definition the closure of X under the above operation.

Remark 5: Let q be a type of elements of a group G.
(a) q is the generic type of some ∞-definable subgroup of G iff q satisfies: for $(a_1,a_2) \models q^2$, $a_1a_2 \models q$.
(b) q is the generic type of a ∞-definable coset of G iff q satisfies: for $(a_1,a_2,a_3) \models q^3$, $a_1a_2^{-1}a_3 \models q$.

Proof:

(a) Let $S = \{ab: a \models q \text{ and } b \models q\}$. Note that if $(a_1,a_2) \models q^2$, then $a_1a_2 \models q \mid a_1$: for any translation-invariant rank, $rk(q) = rk(a_1a_2/\emptyset) \geq rk(a_1a_2/a_1) = rk(a_2/a_1) = rk(a_2/\emptyset) = rk(q)$, so equality holds. It follows that whenever $(b_1b_2) \models q^2$, $b_1^{-1}b_2 \models q$. To see that S is a group, it suffices to show that if a,b,c each realize q, then $abc=de$ for some d, e realizing q. Let $f \models q \mid \{a,b,c\}$. Then $abc = (af)(f^{-1}bc)$. By the previous remark, $f^{-1}b \models q$, and one sees easily that $f^{-1}b \downarrow C$; so $f^{-1}bc \models q$. Let $d = (af)$, $e = (f^{-1}bc)$.

(b) Let $r = stp(a_1a_2^{-1})$ for $(a_1,a_2) \models q^2$. r clearly satifies the condition for being the generic type of a group S. By definition of r, q is the generic type of Sa_2 whenever $a_2 \models q$.

Proposition 6 (Indecomposability): Let G be a group, p a regular type. Let $F^* = \{s: s = stp(a/\emptyset)$ for some $a \in G$, $s \,\square\, p^n$ for some $n > 0$, and s is p-simple$\}$. Assume $w_p(s)$ is bounded for $s \in F^*$. Then:

(a) If $r \in F^*$, then the coset generated by $\{x \in G: x \models r\}$ is ∞-definable, with generic type in F^*.

(b) Let G_i $(i \in I)$ be a collection of ∞-definable connected groups whose generic types are in F^*. Then the group generated by $\cup_i G_i$ is ∞-definable, and its generic type is in F^*.

Proof: Define two operations on F^*: if $q,r \in F^*$, choose $(a,b) \models q \otimes r$, and let $q^{-1} = stp(a^{-1})$, $q \cdot r = stp(a \cdot b)$. Note that $\cdot$ is associative. Also, $w_p(q^{-1}) = w_p(q)$, and $w_p(q \cdot r) \geq w_p(q)$. $[w_p(ab) \geq w_p(ab/b) = w_p(a/b) = w_p(a)]$.

Let F be a subset of F^* closed under the operation: $(q_1,q_2,q_3) \rightarrow q_1 \cdot q_2^{-1} \cdot q_3$. Choose $q \in F$ of largest p-weight. Then for any $r \in F$ one has $w_p(q) \geq w_p(q \cdot r^{-1}) = w_p(q)$. Recall that if t is p-simple and $\square\, p^n$ then every forking extension of t has smaller p-weight than t. It follows that if $(a,b,c) \models q \otimes r \otimes r$ then $ab^{-1} \downarrow b$. Since we also have $ab^{-1} \downarrow c$, $stp(b/ab^{-1}) = stp(c/ab^{-1})$, so there exists $a' \models q$ such that $ab^{-1} = a'c^{-1}$. So $ab^{-1}c = a'c^{-1}c = a'$. This shows that with our choice of q, $q \cdot r^{-1} \cdot r = q$ for any $r \in F$. In particular,

$q \cdot q^{-1} \cdot q = q$, so by Remark 5, q is the generic type of some ∞-definable coset.

To prove (b), let q_i be the generic type of G_i, and let F be the closure of $\{q_i : i \in I\}$ under the operation defined in the first line of the previous paragraph. One obtains an ∞-definable coset C with generic type q such that $q \cdot q_i^{-1} \cdot q_i = q$ for each i. It follows that each $G_i \subset S$, where S is the subgroup of G for which C is a right coset of S. From the construction of C it is clear that S is contained in the subgroup generated by the union of the G_i's (indeed C is). So this subgroup equals S.

For (a), let F be the closure of $\{r\}$ under the same operation. So $F = \{r_n : n \text{ odd}\}$, where $r_n = r \cdot r^{-1} \cdot r \cdot \ldots \cdot r^{\nu(n)}$ (n times; $\nu(n) = (-1)^n$). Let q, C be as in the first paragraph. Then it is clear that C is contained in the coset generated by r. Conversely, say $q = r_n$. From $q \cdot r^{-1} \cdot r = q$ one sees that r_m depends only on the parity of m for $m \geq n$. So $q = r_{2n+1} = r_n \cdot r_n^{-1} \cdot r = q \cdot q^{-1} \cdot r$. From this, one sees easily that every realization of r lies in the coset generated by the realizations of q.

References

[H] Hrushovski, E., Doctoral Dissertation, Berkeley 1986.

[N] Nesin, A., Doctoral Dissertation, Yale 1985.

[M] Makkai, M., A survey of basic stability theory, Isr. J. Math. v. 49 Nos. 1-3 (1981), pp. 181-238.

[Z] Zil'ber, B., Some model theory of simple algebraic groups over algebraically closed fields, Coll. Math. XLVII (1984),pp. 175-180.

ON THE EXISTENCE OF ∅-DEFINABLE NORMAL SUBGROUPS OF A STABLE GROUP

Anand Pillay*

There is a family of results concerning the existence of (∅)-definable normal subgroups of a stable group. Namely:

(1) (Berline-Lascar [B-L]). If G is superstable, and

$$U(G) = \omega^{\alpha_1}n_1+...+\omega^{\alpha_k}n_k + \beta \ (\beta<\omega^{\alpha_k})$$

then G has a normal subgroup K with $U(K) = \omega^{\alpha_1}n_1 + ... + \omega^{\alpha_k}n_k$.

(2) (Hrushovski [H]). If G is stable and its generic type is nonorthogonal to a regular type p, then there is a definable normal subgroup K of G such that G/K is "p-internal" and infinite.

(3) (Pillay-Hrushovski [PH]). If G is 1-based and connected then every type $q = stp(a/A)$ $(a \in G)$ is the generic type of a coset of a normal acl(∅)-definable subgroup K of G.

In this expository paper we will prove these results and some variants.

We work throughout over acl (∅) (i.e. we assume acl (∅) = dcl (∅)). G is assumed throughout to be a <u>saturated</u>, <u>stable</u>, <u>connected</u> group. We prove:

Theorem A. Let g be a generic of G (over ∅), and let $X \subset G$ be invariant and internally closed. Then there is a ∅-definable normal $K \subset G$ such that (i) $G/K \subset X$ and (ii) some stationarisation of tp(g/X) is the generic type of a generic coset of K° (the connected component of K).

---

* Partially supported by NSF Grant DMS 8601289

(We will see that the Berline-Lascar result (1) above follows from Theorem A when we take $X = \{a \in G^{eq}: U(tp(a/\emptyset)) < \omega^{\alpha_k}\}$).

Theorem B. Let g be a generic of G, and $A \subseteq G^{eq}$. Then stp(g/A) is the generic type of a generic coset of a $\emptyset$-definable normal subgroup K of G if and only if for any generic a of G with $a \downarrow_{\emptyset} gA$, both $stp(g \cdot a/aA)$ and $stp(a \cdot g/aA)$ are 1-based types.

First some explanation of the notation:

Definition 1. (a) Let $X \subset M^{eq}$ (M stable saturated). We say X is <u>invariant</u> if X is fixed setwise by any automorphism of M. This clearly means that whether or not $a \in X$ depends on $tp(a/\emptyset)$.
(b) Let $X \subseteq M^{eq}$ be invariant. We say that X is <u>internally closed</u> if for any $a \in M^{eq}$, if for some $b \in M^{eq}$ $a \downarrow b$ and $a \in dcl(X \cup b)$ then also $a \in X$.
(c) $q = stp(a/A)$ is said to be <u>1-based</u> if $Cb(q) \subset acl(a)$.

The next Lemma also appears in [P].

Lemma 2. Let g be a generic of G, and $b \in G^{eq}$. Let $q = tp(g^\wedge b/\emptyset)$. Let $K = \{a \in G$: for $g'^\wedge b'$ realising $q|a$, $tp(a \cdot g'^\wedge b'/a) = q|a.\}$ Then for generic a of G, a/K (the left coset aK), is in the internal closure of $tp(b)^G$.

Proof. Let G_1 be saturated, $G_1 \downarrow_{\emptyset} g^\wedge b$, and let $Y \subset G_1$ be a large independent set of generics. Let X = the set of realisations of tp(b) in G_1^{eq}. Let $a \in G_1$ be generic and independent with Y over $\emptyset$. Note that

$$tp(a \cdot g/G_1) \text{ is generic.} \qquad (*)$$

Claim. $tp(a \cdot g^\wedge b/G_1)$ is definable over $Y \cup X$.
Proof of Claim. It is enough to show that $tp(a \cdot g^\wedge b/G_1)$ is finitely satisfiable in $Y \cup X$. So suppose $\bar{m} \subset G$, and $\models \varphi(a \cdot g, b, \bar{m})$. As Y is large there is $c \in Y$ with $\bar{m} \downarrow_{\emptyset} c$. By (*) $tp(a \cdot g/\bar{m}) = tp(c/\bar{m})$. Thus $\models$

$\exists y(\varphi(c,y,\bar{m}) \wedge r(y))$ where $r = tp(b/\emptyset)$. As G_1 is saturated, we can find such a y in G_1, i.e. in X. This proves the claim.

Let $Y_o \cup X_o \subset Y \cup X$ be small such that $tp(a \cdot g\char`^b/G_1)$ is definable over $Y_o \cup X_o$. As $tp(g\char`^b/G_1)$ is definable over $\emptyset$ we see that any automorphism f of G_1 takes $tp(a \cdot g\char`^b/G_1)$ to $tp(f(a) \cdot g\char`^b/G_1)$. Thus, if f is a $Y_o \cup X_o$ automorphism of G_1, then $tp(a \cdot g\char`^b/G_1) = tp(f(a) \cdot g\char`^b/G_1)$, i.e. $tp(g\char`^b/G_1) = tp(a^{-1} f(a) \cdot g\char`^b/G_1)$, that is $a^{-1} \cdot f(a) \in K$. Thus $a^{-1} \cdot f(a) \in K$. Thus $a/K \in dcl(Y_o \cup X_o)$. But K is $\emptyset$-definable and $a \downarrow Y_o$. Thus a/K is in the internal closure of X as required. □

Lemma 3. Let $X \subseteq G^{eq}$ be invariant and internally closed. Let K be $\emptyset$-definable such that for generic a of G, a/K (left coset) is in X. Let L = intersection of conjugates K^g of K, $g \in G$. Then $G/L \subset X$ (and L is normal $\emptyset$-definable).

Proof. First $L = K \cap K^{g_1} \cap \ldots \cap K^{g_n}$. Clearly L is normal and $\emptyset$-definable. Note that if a is a generic of G over g_i then a/K^{g_i} is interdefinable with a^{g_i}/K over g_i, and moreover a^{g_i}/K is a generic coset of K, and thus is in X. Thus, if a is a generic of G over $g_1,\ldots,g_n$ then $a/L \in dcl(a/K, \ldots, a^{g_i}/K, g_1,\ldots,g_n)$. So a/L is in A (as X is internally closed) As every element of G/L is a product of generics of $G/L \subset X$. □

Proof of Theorem A

Let K be the intersection of all $\emptyset$-definable subgroups L of G such that for generic a of G, $a/L \in X$. By Lemma 3, K is normal, clearly $\emptyset$-definable, and $G/K \subset X$. (K need not be connected). Let a realise the generic of K° over G.

So by definition, $tp(a \cdot g/G)$ is a generic of the coset g/K° $(=a \cdot g/K^\circ)$ of K° and so is definable over g/K°. On the other hand $g/_K \in X$ and $g/K^\circ \in acl(g/K)$. Thus $tp(a \cdot d/G)$ does not fork over X.

But by Lemma 2, $tp(a \cdot g/X) = tp(g/X)$. Thus $tp(a \cdot g/G)$ is a stationarisation of $tp(g/X)$ and we finish.

Before proving the Berline-Lascar theorem, we recall some facts about U-rank.

Fact 4. $U(a/bA) + U(b/A) \leq U(a\hat{}b / A) \leq U(a/bA) \oplus U(b/A)$, and of course $U(a\hat{}b/A) = U(b\hat{}a/A)$.

Lemma 5. Let $U(a/A) = \omega^{\alpha_1}n_1 + ... + \omega^{\alpha_k}n_1 + \beta$ where $\alpha_1 > \alpha_2 > ... > \alpha_k$ and $\beta < \omega^{\alpha_k}$. Let $B \supset A$, and $U(a/B) = \omega^{\alpha_1}n_1 + ... + \omega^{\alpha_k}n_k$; let $c \in Cb(stp(a/B))$. Then $U(c/A) < \omega^{\alpha_k}$.

Proof. Let $a_1, a_2, ...$ be a Morley sequence in $stp(a/B)$. So $c \in acl(a_1, ..., a_n)$ for some $n < \omega$. By Fact (4) we have: ($\bar{a}$ denotes $(a_1, ..., a_n)$)
$U(\bar{a}/cA) + U(c/A) \leq U(\bar{a}c/A) \leq U(c/\bar{a}A) \oplus U(\bar{a}/A)$.
But $\omega^{\alpha_1}n_1 \cdot n + ... + \omega^{\alpha_k}n_k \cdot n \leq U(\bar{a}/cA)$, $U(c/\bar{a}A) = 0$ and $U(\bar{a}/A) < \omega^{\alpha_1}n_1 \cdot n + ... + \omega^{\alpha_k}(n_k \cdot n + 1)$.
Thus $\omega^{\alpha_1}n_1 \cdot n + ... + \omega^{\alpha_k}n_k \cdot n + U(c/A) < \omega^{\alpha_1}n_1 \cdot n + ... + \omega^{\alpha_k}(n_k \cdot n + 1)$, whereby $U(c/A) < \omega^{\alpha_k}$. □

Corollary 6. Let $U(G) = \omega^{\alpha_1}n_1 + ... + \omega^{\alpha_k}n_k + \beta$ ($\beta < \omega^{\alpha_k}$). Then there is a normal $\varnothing$-definable subgroup K of G, with $U(K) = \omega^{\alpha_1}n_1 + ... + \omega^{\alpha_k}n_k$.

Proof. By Lemma 5, there is $c \in G^{eq}$ with $U(c/\varnothing) < \omega^{\alpha_k}$ and $U(g/c) = \omega^{\alpha_1}n_1 + ... + \omega^{\alpha_k}n_k$ (where g is a generic of G over $\varnothing$). On the other hand, if $X = \{a \in G^{eq}: U(a) < \omega^{\alpha}\}$, then by Fact 4 $U(g/X) \geq \omega^{\alpha_1}n_1 + ... + \omega^{\alpha_k}n_k$. Thus $U(g/X) = \omega^{\alpha_1}n_1 + ... + \omega^{\alpha_k}n_k$. Clearly X is invariant and internally closed. So by, Theorem A we find a subgroup K of G which is normal and $\varnothing$-definable, with $U(K) = \omega^{\alpha_1}n_1 + ... + \omega^{\alpha_k}n_k$ ($tp(g/X)$ is the generic type of a coset of K, so $U(g/X) = U(K)$). □

We now consider Hrushovski's result. Let $p \in S(A)$ be a regular type, not orthogonal to $\emptyset$. We say c is p-internal over $\emptyset$ if $\exists$ B $c \downarrow_{\emptyset} B$ and realisations $d_1 \dots d_k$ of extension of conjugates of p over B such that $c \in$ dcl $(B, d_1, \dots, d_k)$. Note that $\{c : c$ is p-internal over $\emptyset\}$ is invariant and internally closed.

Lemma 7. (T stable). Let tp(a/∅) be nonorthogonal to p (p regular). Then there is c which is p-internal over $\emptyset$ such that $a \not\downarrow c$. (In fact we can choose $c \in$ dcl(a), and c having "nonzero p-weight").

Proof Let $a \downarrow B$ and let e realise p|B, such that $a \not\downarrow_B e$. Let D = Cb(stp(eB/a)). So $D \not\subseteq$ acl (∅). Let $\{e_iB_i : i < \omega\}$ be a Morley sequence in stp(eB/a). Then $D \subseteq$ dcl $(\{e_i, B_i : i < \omega\})$, and $D \subseteq$ acl (a).
A nonalgebraic member of D will then satisfy the requirements (as $D \subset$ acl(a), $D \downarrow_{\emptyset} \{B_i : i < \omega\}$). □

Now we can obtain Hrushovski's result (2) mentioned in the introduction: for suppose the generic type of G is nonorthogonal to regular p. By Lemma 7, we can find generic g of G and c p-internal over $\emptyset$ such that $g \not\downarrow c$. Let $X = \{c \in G^{eq}$: c is p-internal over $\emptyset\}$. By Theorem A, there is ∅-definable normal $K < G$ with $G/K \subset X$. (In fact it turns out that the generic of G/K again has "nonzero p-weight").

Finally, by a slight refinement of [PH] we give necessary and sufficient conditions for stp(g/A) (g generic of G) to be a generic type of a (generic) coset of an ∅-definable normal subgroup.

Proof of Theorem B.

Suppose first stp(b/A) to be generic type of a generic coset of K, where K is ∅-definable, normal and of course connected. Let a be generic of G over g^A. So stp(g/aA) does not fork over A. Let $G_1 \supset a A$, stp(g/G_1) dnf over A. Let q = generic type of K over G_1. By assumption,

$tp(g/G_1) = q \cdot b = b \cdot q$ for some $b \in G$. But then $tp(g \cdot a/G_1) = tp\ (q \cdot (b \cdot a)/\ G_1)$ and $tp(a \cdot g/G_1) = tp\ ((a \cdot b) \cdot q/G_1)$. But then $tp\ (g \cdot a/G_1)$ is definable over $b \cdot a/K = g \cdot a/K \in dcl(g \cdot a)$, so is 1-based. Similarly $tp(a \cdot g/G_1)$ is 1-based.

Conversely, suppose the right hand side conditions hold. Again, let a be generic of G over gA, and let $G_1 \supset A \cup a$ with $tp(g/G_1)$ not forking over A.

Let K_1 = left stabiliser of $tp(g/G_1)$ (= $\{b \in G_1 : tp(b \cdot g/G_1) = tp\ (g/G_1)\}$), and let K_2 = right stabiliser of $tp(g/G_1)$. So K_1, K_2 are both acl(A)-definable.

Then any automorphism of G_1 which fixes acl(A) and $tp(a \cdot g/G_1)$ fixes a/K_1 (=left coset aK_1). Thus as $tp(a \cdot g/G_1)$ is based, $a/K_1 \in acl(A \cup a \cdot g)$. On the other hand $stp(g/a \cdot g \cup a/K_1 \cup A)$ does not fork over A. But the right coset of g mod K_1 is definable over $a \cdot g \cup a/K_1 \cup acl(A)$.

Thus the right coset $g/K_1 \in stp(g/a \cdot g \cup a/K_1)$. It easily follows that $stp(g/A)$ is the generic type of the right coset $K_1 \cdot g$ of K_1. (1)

We can do the same thing for K_2, deducing that $stp(g/A)$ is the generic type of the left coset $g \cdot K_2$ of K_2. (2)

Note also that K_1 = left stabiliser of $tp(g \cdot a/G_1)$, so as the latter type is by assumption 1-based, K_1 is $acl(g \cdot a)$-definable. (3)

Similarly K_2 is $acl(a \cdot g)$-definable. (4)

As $g \cdot a \downarrow A$, $a \cdot g \downarrow A$, it follows from (3),(4) and the acl(A)-definability of K_1, K_2, that both K_1, K_2 are $acl(\varnothing)$-definable.

It easily follows from (1) and (2) that $K_1^g = K_2$. As both K_1, K_2 are acl $(\varnothing)$-definable and g is generic over $\varnothing$ it follows that for generic independent g_1, g_2 of G $K_1^{g_1} = K_2 = K_1^{g_2}$, and $K_1^{g_1 g_2^{-1}} = K_1$.

But $g_1 \cdot g_2^{-1}$ is also generic, so $K_1^g = K_1$ for generic g. Thus K_1 is normal and $K_1 = K_2$, proving the Theorem.

References

[B-L]. Chantale Berline and Daniel Lascar, Superstable groups, Annals of Pure and Applied Logic, 30 (1986).

[H]. E. Hrushovski, Ph.D. Thesis, Contributions to Stable Model Theory, Berkeley 1986.

[P-H]. A. Pillay and E. Hrushovski, Weakly normal groups. Logic Colloquium '85, North-Holland 1987.

[P]. Bruno Poizat, Groupes stables, Nur Al - Mantiq Wal - Ma'rifah, 1987.

THE GROUP CONFIGURATION – after E. Hrushovski

Elisabeth Bouscaren

We present here some results of E. Hrushovski which give, in the context of stable theories, an "abstract" or geometrical (in terms of dependence relations), characterization of the presence of some group acting definably on a weight one type.

Preliminaries

We will use freely definitions and basic facts concerning local weight (i.e. p–weight, for p a given regular type), as introduced in [Hr1]; these can also be found in [Hr2] or [Po].

We just introduce the following definition:

Definition: Let p be a fixed regular type (over $\varnothing$) and let $\bar{a}$,$\bar{b}$ be such that $t(\bar{a}/A)$ and $t(\bar{b}/A)$ are p–simple. We say that $\bar{a}$ and $\bar{b}$ are <u>p–independent</u> over A (denoted $\bar{a} \perp_p \bar{b}$) if $w_p(\bar{a}\bar{b}/A = w_p(\bar{a}/A) + w_p(\bar{b}/A)$.

We need to recall briefly what is meant by the canonical basis of a non stationary type.

We begin with the following definitions and theorems which can be found in [Ls., Chapt. 3–2] or in [Pi., Chapt. 4].

Definition:

Let T be a stable theory, $p \in S(A)$. A definition of p is a map d, which takes each formula $\varphi(\bar{v},\bar{y})$ to a formula $d_\varphi(\bar{y})$ such that:

i) for all $\bar{a} \in A$, p $\vdash \varphi(\bar{v},\bar{a})$ iff $\models d_\varphi(\bar{a})$

ii) for all $B \supseteq A$, for all formulas $\varphi(\overline{v},\overline{y})$, and for all $\overline{b} \in B$, $\models d_\varphi(\overline{b})$ iff all non forking extensions of p over B satisfy the formula $\varphi(\overline{v},\overline{b})$. We say that $p \in S(A)$ is <u>definable over</u> A_0 if there is a map d satisfying conditions (i) and (ii) such that, for all $\varphi(\overline{v},\overline{y})$, the formula $d_\varphi(\overline{y})$ has its parameters in A_0.

Weak definability theorem:

Let T be a stable theory, $p \in S(A)$, then p is definable over A. In fact, there is $A_0 \subseteq A$, $|A_0| \leq |L| + \aleph_0$ such that p is definable over A_0.

Theorem:

Let T be a stable theory, $A \subseteq B$, $p \in S(B)$; p is definable over A if and only if p is the unique non forking extension in S(B) of its restriction to A.

Notation – dcl(A) is the definable closure of A
– acl(A) is the algebraic closure of A.

Recall that we are working inside a big saturated model of T, $\mathfrak{C}$. The following theorem is the analogue of Theorem 6–10, Chapt. III–6 in [Sh.], where the canonical basis is defined for a stationary type.

Theorem: Existence of the canonical basis

Let T be a stable theory; in T^{eq}, for every type $p \in S(A)$, there is a set C(p) such that

(i) $C(p) \subseteq dcl(A)$; p is definable over C(p)

(ii) an automorphism σ of $\mathfrak{C}^{eq}$ leaves C(p) pointwise fixed if and only if σ leaves the set of non forking extensions of p globally invariant

(iii) $D \subseteq A$ is such that p does not fork over D, if and only if $C(p) \subseteq acl(D)$.

(We say that C(p) is the canonical basis of p.)

In the case of a stationary type q, the canonical basis of q is contained in the definable closure of any Morley sequence in q. The analoque in the non stationary case is:

Fact: Let T be a stable theory, $p \in S(A)$ and Let I be an independent set of realizations of p over A, containing a Morley sequence in each strong type extending p. Then C(p) is contained in the definable closure of I.

Configurations

We will assume that we are working with T a complete superstable theory, but the theorem is in fact true for any stable theory, with only minor changes in the proofs.

Let p be a fixed regular type (over $\varnothing$).

Definition: The set $\{a_1,a_2,a_3,b_1,b_2,b_3\}$ is called a p–configuration over A if it satisfies the following:

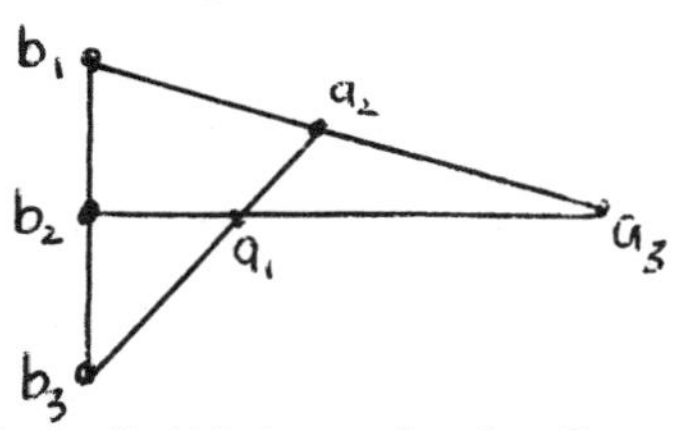

(a) For each i, $t(b_i/A)$ is p–simple of p–weight n and $t(a_i/A)$ is p–simple of p–weight 1.

(b) All elements are pairwise p–independent over A.

(c) $w_p(b_1b_2b_3/A) = 2n$; for all $i \neq j \neq k$, $w_p(b_ia_ja_k/A) = n + 1$; $w_p(\overline{a}\overline{b}/A) = 2n + 1$.

Remarks:

– With the assumptions in (b), (c) implies that b_k is in the p–closure of Ab_ib_j, and that a_k is in the p–closure of Ab_ia_j.

– If any element x of the configuration is replaced by some p–simple element y such that $cl_p(Ax) = l_p(Ay)$, then we still have a p–configuration over A.

Main Theorem:

Suppose there is a p–configuration in some model of T. Then there is an ∞ definable group G and a definable generic action of G on a regular type domination equivalent to p.

Lemma 1:

There is a p–configuration $\bar{a},\bar{b}$ over some model M such that $t(\bar{a}\bar{b}/M)$ is equivalent (i.e. domination equivalent) to some power of p (in fact p^{2n+1}).

Proof: First, replace A (which we can suppose to be finite) by an $\aleph_\varepsilon$ – saturated model $M' \supseteq A$, such that the configuration $\bar{a}\bar{b} \perp M'$, then of course we now have a configuration over M'.

Now let N be $\aleph_\varepsilon$ –prime over $M'\bar{a}\bar{b}$ and let M, $M' \leq M \leq N$ be maximal orthogonal to p over M'. Then in N, if $t(e/M')$ is p–simple of p–weight 0, $e \in M$: if not, Me contradicts the maximality of M. If $e \in N$ and $t(e/M')$ is p–simple of weight $n > o$, then $t(e/M)$ Δ p^n, more precisely, there is $\bar{\alpha}$ in N realizing p^n such that $\bar{\alpha}\, \Delta\, e$ over M' and $e\, \Delta\, \bar{\alpha}$ over M: as $t(e/M')$ is p–simple, we know that there is $\bar{\alpha}\beta$ in N such that $\bar{\alpha}$ realizes p^n and $t(\beta/M')$ is hereditary orthogonal to p, and e and $\bar{\alpha}\beta$ are domination equivalent over M'. Then β must be in M and the rest follows. It is now easy to check that $\bar{a}\bar{b}$ is a p–configuration over M. □

Remark: Note that now, in N, for p–simple elements, p–independence over M is equivalent to independence over M.

Lemma 2:

There is, in N, a p–configuration over M such that $a_1 \in \mathrm{dcl}\,(Ma_3b_2)$ and $a_2 \in \mathrm{dcl}(Ma_3b_1)$.

Proof: We will replace the given configuration by one where

$a_1 \in \mathrm{dcl}\,(Ma_3b_2)$, from this one, we then get exactly in the same way, another one satisfying also the second condition.

Let $C \subseteq M$, finite, be such that $t(\overline{ab}/M)$ is based on C (i.e. $t(\overline{ab}/M)$ does not fork over C and $t(\overline{ab}/C)$ is stationary) and let $b'_1 \in M$ have the same type as b_1 over C. Then, as b_1 and $b_2\, a_1\, a_3$ are independent over C $b_1\, b_2\, a_1\, a_3$ and $b'_1\, b_2\, a_1\, a_3$ have the same type over C; let N be the $\aleph_\varepsilon$-prime model over $M\overline{ab}$ and let b'_3 and a'_2 in N be such $b_1\, b_2\, b_3\, a_1\, a_2\, a_3$ and $b'_1\, b_2\, b'_3\, a_1\, a'_2\, a_3$ have the same type over C. It follows that $cl_p(Mb_2) = cl_p(Mb_2b'_3)$: by condition (c), we have that b'_3 is in the p–closure of Cb'_1b_2, so $b'_1 \in cl_p(Mb_2)$. Similarly, a_3 and $a_3\, a'_2$ have the same p–closure over M.

Now let a'_1 be the canonical basis of $t(a_1/Mb_2b'_1a_3a'_2)$ (in the sense described above in the preliminaries) or more precisely a finite subset of the basis over which it is algebraic.

Now we will see that a'_1 and a_1 have the same p–closure over M: let e be any element realizing the same type as a_1 over $Mb_2b'_3a_3a'_2$. Now,

$$w_p(a_1eb_2b'_3a_3a'_2/C) = w_p(a_1e/C) + w_p(b_2a_3/Ca_1e) + w_p(b'_3a'_2/Ca_1eb_2a_3).$$

As both a_1 and e are in the p–closure of Cb_2a_3,

$$w_p(a_1eb_2a_3/C) = w_p(b_2a_3/C) = n+1 = w_p(a_1e/C) + w_p(b_2a_3/Ca_1e).$$

We know that $w_p(a_1e/C) \leq 2$; suppose it were equal to 2, then it would follow that $w_p(b_2a_3/Ca_1e) = n-1$. It would follow as easily that $w_p(b'_3a'_2/Ca_1eb_2a_3)$ must be at most $n-1$. But this contradicts the fact that, as $b'_1 \in cl_p\,(Cb_2b'_3)$, $w_p\,(a_1eb_2b'_3a_3a'_2/C)$ must be $2n+1$. Hence $w_p(a_1e/M) \leq w_p(a_1e/C) = 1$, that is, $e \in cl_p\,(Ma_1)$.

Now let I be an independent set of realizations of $t(a_1/Mb_2b'_3a_2a'_3)$ such that $a'_1 \in \mathrm{dcl}\, MI$. By the above I is in the p–closure of Ma_1, hence so is a'_1. As a_1 has p–weight 1 over M, and a'_1 must have p–weight > 0 over M it follows that $cl_p(Ma_1) = cl_p(Ma'_1)$.

So, as remarked at the beginning, if we replace b_2 by $b_2b'_3$ a_3 by a'_2a_3, and a_1 by a'_1, we still have a p–configuration over M. Now a'_1 is in the canonical basis of $t(a_1/Mb_2b'_3a_3a'_2)$, hence it is in the definable closure of $b_2b'_3a_3a'_2$ over M. □

Lemma 3:

There is a p–configuration over M such that $a_3 \in dcl\,(Ma_1b_2) \cap dcl\,(Ma_2b_1)$, and which still satisfies the conditions in lemmas 1 and 2.

Proof: Let $\overline{ab}$ be the configuration over M given by the preceeding lemma. Let D be the canonical basis of $t(a_3/Mb_1b_2a_1a_2)$. Then $a_1 \in dcl(MDb_2)$: By the properties of D mentioned in the preliminaries $t(a_3/D)$ has a unique nonforking extension over MDb_2a_1. Now let a'_1 be such that $t(a'_1/MDb_2) = t(a_1/MDb_2)$, and let a''_3 realise a nonforking extension of $t(a_3/D)$ over $MDb_2a_1a'_1$. By what we have just said $t(a''_3a'_1b_2DM), = t(a''_3a'_1b_2DM)$, $(=t(a_3a_1b_2DM))$, so as $a_1 \in dcl(Ma_3b_2)$, we see that $a_1 = a'_1$. For the same reason, $a_2 \in dcl\,(MDb_1)$ also. Now let a'_3 be a finite subset of D such that D is in the algebraic closure of a'_3 and both a_1 and a_2 are definable over Ma'_3b_2, Ma'_3b_1 respectively. Arguing similarly as we did in Lemma 2 for a'_1, we see that a_3 and a'_3 have the same p–closure over M. We can therefore replace a_3 by a'_3 without any loss and we now have a configuration above M satisfying all the preceding conditions and the added one that $a_3 \in dcl\,(Mb_1b_2a_1a_2)$.

Let $C \subseteq M$, finite be such that $t(\overline{a}\overline{b}/M)$ is based over C. Now let this time $b'_3 \in M$ have same type as b_3 over C. As in lemma 2, find b'_1 and a'_2 in N, the $\aleph_\varepsilon$–prime model over $M\overline{a}\overline{b}$, such that $b'_1b_2b'_3a_1a'_2a_3$ is a configuration isomorphic to the original one over C, find also b'_2 and a'_1 such that $b_1b'_2b'_3a'_1a_2a_3$ is also isomorphic to the original configuration over C.

We want to replace, in our original configuration, b_1 by $b_1b'_2$, b_2 by $b_2b'_1$, a_1 by $a_1a'_2$ and a_2 by $a_2a'_1$. We must check that the new elements have the same p–closures as the old ones over M. As in lemma 2, this follows directly from condition (c) in a configuration.

Let us check now that this new configuration over M satisfies all the requirements:

–$a_1 \in dcl(Ma_3b_2)$ and $a'_2 \in dcl(Ma_3b'_1)$, so $a_1a'_2 \in dcl(Ma_3b_2b'_1)$

–$a_2 \in dcl(Ma_3b_1)$ and $a'_1 \in dcl(Ma_3b'_2)$, so $a_2a'_1 \in dcl(Ma_3b_1b'_2)$

–by isomorphism also, as this was true for a_3 in the original configuration, we have that $a_3 \in dcl(Ma_1a'_2b'_1b_2)$ and $a_3 \in dcl(Ma'_1a_2b_1b'_2)$. □

By lemmas 2 and 3 and the definability relations they give, let $a_3 = f_{b_1}(a_2)$ and $a_1 = g_{b_2}(a_3)$, where f_b is an invertible Mb_1–definable function, and g_{b_2} is an Mb_2–definable invertible function.

Let us denote by q_i the type of a_i over M.

Let $h_{b_1b_2}(a_2)$ denote $g_{b_2}(f_{b_1}(a_2))$. Define the germ of $h_{b_1b_2}$ as the equivalence class of b_1b_2 modulo the (definable) relation b_1b_2 and $b'_1\, b'_2$ are the equivalent if for all a realizing q_2, independent from $b_1b_2b'_1b'_2$, $h_{b_1b_2}(a) = h_{b'_1b'_2}(a)$.

Lemma 4:

The germ of $h_{b_1b_2}$ is in the p–closure of Mb_3.

Proof: Indeed, $h_{b_1b_2}(a_2) = a_1$ is in $cl_p(Ma_2b_3)$: as a_2 is independent from $\bar{b}$ ($\bar{b} = b_1b_2$) over M, for any a realizing q_2 and independent from $\bar{b}$, we also have that $h_{b_1b_2}(a) \in cl_p(Mab_2)$. Let $(e_i)_{i\in I}$ be a Morley sequence of q_2 over $M\bar{b}$. Then the germ of $h_{b_1b_2}$ is definable over $\{(e_i)_{i\in I},(h_{b_1b_2}(e_i))_{i\in I},\} \subseteq cl_p\{M,b_3,(e_i)_{i\in I},\}$. Now as $b_1b_2 \perp_p (e_i)_{i\in I}$, over Mb_3, so is the germ, and in fact, it must be in $cl_p(Mb_3)$. □

Define F to be the set of germs of $\{f_b;\ t(b/M) = t(b_1/M)\}$ and let G be the set of germs of $\{g_e;\ t(e/M) = t(b_2/M)\}$.

From now on it is understood that we are always working over the model M, and that we are always considering germs of functions, even if we do not mention it explicitly.

Lemma 5:

(1) If $g \in G$, $f \in F$, f and g are independent, then $g\circ f$ and f, and $g\circ f$ and g are independent.

(2) If f, f' are independent in F, then $h = f^{-1}\circ f'$ is independent from f and f' and is q_2-internal.

(3) If $f_o,f_1,f_2,f_3 \in F$ and are independent, then there are $f_5,f_6 \in F$, independent, such that $f_o^{-1}\circ f_1 \circ f_2^{-1}\circ f_3 = f_5^{-1}\circ f_6$, and $f_5^{-1}\circ f_6$ is independent from both $f_o^{-1}\circ f_1$ and $f_2^{-1}\circ f_3$.

Proof:

(1) g corresponds to some b_2 and f to some b_1, then by lemma 4, $g\circ f$ is in the p-closure of b_3, where b_3 and b_1 and b_3 and b_2 are independent.

(2) Let f be f_{b_1} and f' be $f_{b_1'}$ with b_1 and b_1' independent over a_3. Then let $f_{b_1}(a_2') = a_3$ and $f^{-1}(a_3) = a_2$, so $h(a_2') = a_2$. Now h is definable over a Morley sequence e_i in type a_2' over b_1b_1' together with the $h(e_i)$'s, so it is q_2-internal. Now the sequence $e_i\, h(e_i)$ is independent from b_1 and from b_1' hence so is h.

(3) Let $g \in G$ be independent from $\{f_o,f_1,f_2,f_3\}$. Note first that $(g\circ f_o)^{-1}$, $(g\circ f_1)$, $(f_o^{-1}\circ f_1)$ are pairwise independent:

g corresponds to some b_2, f_o to some b_1 and f_1 to some b_1', such that $t(b_1b_2) = t(b_1'b_2)$. Let b_3' complete the diagram, i.e. be such that $b_1b_2b_3$ and $b_1')b_2b_3'$ have the same type. Then $(g\circ f_o)^{-1} \in cl_p(b_3)$, $(g\circ f_1) \in cl_p(b_3')$, $(f_o^{-1}\circ f_1) \in cl_p(b_1b_1'))$. The result follows by the independence relations.

Now let us write $(f_o^{-1}\circ f_1) = (g\circ f_o)^{-1} \circ (g\circ f_1)$. By independence, $(f_o^{-1}\circ f_1)$, $(g\circ f_o)^{-1}$ and $(f_2^{-1}\circ f_3)$, $(g\circ f_1)^{-1}$ have the same type, hence there is some h of the form $(g'\circ f')$ such that $(f_2^{-1}\circ f_3) = (g\circ f_1)^{-1}\circ h$. Now $(g\circ f_o)^{-1}$, h have the same type as $(g\circ f_1)^{-1}$, h, therefore $(g\circ f_o)^{-1}\circ h$ is equal to some $(f_5^{-1}\circ f_6)$ and the result follows. Now, $(f_5^{-1}\circ f_6)$ and $(f_2^{-1}\circ f_3)$ are equidefinable over $(f_o^{-1}\circ f_1)$, hence have same p-weight; the independence follows.

Now let $H = \{f^{-1}\circ f';\ f, f'$ independent in $F\}$. Then H is closed under inverse, and by the above lemma it is closed under generic composition. Now let $G = \{h\circ h';\ h, h' \in H\}$. Then G is closed under inverse, and G is also closed under composition: let $h, h', h'' \in H$, we see that $h\circ h'\circ h'' \in G$: choose g in H independent from all the rest, then by the above lemma, there is g' in H, g' independent from g, such that $h' = g\circ g'$ and g' independent from h'. So $h\circ h' h'' = h\circ g\circ g'\circ h''$, now $h\circ g \in H$ because we chose g independent from h; $g \underset{h'}{\perp} h''$ hence $g' \underset{h'}{\perp} h''$, $g'\circ h'' \in H$. This finishes the proof of the main theorem. □

So we have the group G acting generically on the type q_2.

Proposition 1:

The group G is connected, with generic any element of H, that is elements of the form $f^{-1} \circ f'$ where f and $f' \in F$ are independent. G is q_2-internal.

Proof: In order to see that elements of H realize the unique generic of G, we check that for $g \in G$ and $h \in H$ independent, $g \circ h \in H$ and is independent from g. Let $g = h_1 \circ h_2$, wlog, suppose that $h \perp h_1 h_2$. Now $g \circ h = h_1 \circ (h_2 \circ h)$, $h_2 \circ h \in H$ and is independent from h_1, so this again is the product of two independent elements of H and the rest follows directly from lemma 5.

Lemma 6:

If $h \in H$, then the p-weight of h is equal to the p-weight of F.

Proof: h is of the form $f \circ f'$, with f, f' in F, independent, and $h \perp f$, $h \perp f'$. As h and f are equidefinable over f', they must have the same p-weight over f', and both h and f are independent from f'. □

Definition: We say that a p-configuration is <u>minimal</u> if for all i, if there is some $b_i' \in cl_p(b_i)$ such that we still have a p-configuration if we replace b_i by b_i', then $cl_p(b_i') = cl_p(b_i)$.

Proposition 2:

If we start from a minimal p-configuration, then the p-weight of G must be equal to the p-weight of the b_i's.

Proof: Note first that all the changes we have made in lemmas 1, 2, and 3, in order to replace the conditions of p-dependence by conditions of definability, maintain the condition of minimality. Now if we replace in our configuration b_1 by f_{b_1}, b_2 by g_{b_2} and b_3 by $g_{b_2} \circ f_{b_1}$, then we still have a p-configuration. By minimality, then the p-weights of b_1 and f_{b_1} must be the same, and the result follows by the above lemma. □

We have a "generic" transitive action of G on a type q of p-weight 1. By results in [Hr1], we then get a transitive action of a group of p-weight n on an infinitely definable set S, with q generic type of S for the action. The possibilities for such an action are known [Hr1]:

There are only 3 cases:

- n = 1: the group G is abelian and the action is simply transitive.
- n = 2: S is the affine line over a definable field K and G acts as AGL_1 (K).
- n = 3: S is the projective line over a definable field K and G acts as PGL_1(K).

It is not possible to have $n \geq 4$.

References

[Hr1] E. Hrushovski, Ph.D. thesis, Univ. of California, Berkeley, 1986.

[Hr2] E. Hrushovski, Locally modular regular types, Classification Theory (Proceedings, Chicago, 1985), Springer-Verlag, Berlin, 1988.

[Ls] D. Lascar, Stability in Model Theory, Longman Sc. Tech., New York, 1987.

[Pi] A. Pillay, An introduction to Stability Theory, Clarendon Press, Oxford, 1983.

[Po] B. Poizat, Groupes Stables, Nur al-mantiq wal-ma'rifah, Paris, 1987.